BEYOND THE SOLAR SYSTEM

ASTRONOMY LIBRARY NO. 2

By David J. Eicher

Preface 3	NGC 2841 22	The S Nebula 55
Getting to Know the Sky 4	NGC 2903 22	The Butterfly Cluster 55
47 Tucanae 5	M81 22	M7 55
The Andromeda Galaxy 5	M82 23	The Trifid Nebula 56
NGC 253 6	The Ghost of Jupiter 23	The Lagoon Nebula 56
The Magellanic Clouds 6	IC 2602 23	The Parrot's Head Nebula 58
NGC 362 8	The Eta Carinae Nebula 24	70 Ophiuchi 58
The Owl Cluster 8	The Owl Nebula 24	The Small Sagittarius Star Cloud 59
M33 8	M65 25	Barnard 92 59
The Little Dumbbell Nebula 9	M99 25	The Eagle Nebula 61
Gamma Andromedae 9	M106 26	The Omega Nebula 61
The Double Cluster 10	M100 26	M22 62
Mira 10	The Coma Star Cluster 27	R Scuti 62
Algol 11	NGC 4565 27	The Wild Duck Cluster 62
NGC 1360 11	The Sombrero Galaxy 28	The Ring Nebula 63
The Pleiades 12	Porrima 28	NGC 6781 63
The California Nebula 12	M94 28	Albireo 64
The Hyades 13	The Coal Sack 29	The Blinking Planetary 64
The Crab Nebula 13	The Jewel Box Cluster 30	M71 64
The Orion Nebula 14	The Blackeye Galaxy 30	The Dumbbell Nebula 65
The Tarantula Nebula 14	M63 31	NGC 6888 65
The Horsehead Nebula 14	Mizar 31	The Veil Nebula 66
M37 16	Centaurus A 32	The North America Nebula 68
M35 16	Omega Centauri 32	M15 68
The Rosette Nebula 16	The Whirlpool Galaxy 49	M39 68
Hubble's Variable Nebula 18	M83 50	IC 1396 70
The Cone Nebula 18	M3 50	SS Cygni 70
Sirius 19	M101 51	The Cocoon Nebula 71
M41 19	Xi Bootis 51	The Helical Nebula 72
The Eskimo Nebula 19	M5 52	NGC 7331 74
NGC 2403 20	R Coronae Borealis 52	The Bubble Nebula 74
M46 20	M4 53	NGC 7789 75
NGC 2516 21	The Hercules Cluster 53	Photo credits 76
Zeta Cancri 21	NGC 6231 54	Index 78
The Beehive Cluster 21	M92 54	Bibliography 80

Books by David J. Eicher

The Universe from Your Backyard; a guide to deep-sky objects from ASTRONOMY magazine (Cambridge University Press and Kalmbach Publishing Co., New York, 1988)

Deep Sky Observing with Small Telescopes; a guide and reference
(Editor and coauthor; Enslow Publishers, Hillside, New Jersey, 1989)

Stars and Galaxies; ASTRONOMY's guide to exploring the cosmos
(Editor and coauthor; AstroMedia, Waukesha Wisconsin, 1992)

Beyond the Solar System; 100 best deep-sky objects for amateur astronomers, (AstroMedia, Waukesha, Wisconsin, 1992)

FOR LYNDA EICHER, a city girl who has come to know the bright stars of the country

Art Director: Lawrence Luser
Artists: Lisa Bergman, Mark Watson

© 1992 by AstroMedia. All rights reserved. This book may not be reproduced in part or in whole without written permission from the publisher, except in the case of brief quotations used in reviews. Published by AstroMedia, a division of Kalmbach Publishing Co., 21027 Crossroads Circle, P. O. Box 1612, Waukesha, WI 53187. Printed in Hong Kong

Library of Congress Cataloging-in-Publication Data

Eicher, David J., 1961-
  Beyond the solar system: 100 best deep-sky objects for amateur astronomers / David J. Eicher.
    p. cm.
  ISBN 0-913135-10-0
    1. Astronomy—Handbooks, manuals, etc. 2. Astronomy—Observer's manuals. I.Title.
QB64.E52   1992   91-62556
523—dc20   CIP

Preface

As a kid you probably heard about the Orion Nebula and the Andromeda Galaxy and a few other astronomical objects. When I was young I was puzzled by the peculiar inhabitants of the night sky and their strange names. This book is intended to introduce you to the most spectacular wonders of the sky, to erase any lingering questions about what they are. *Beyond the Solar System* is an introduction to the finest celestial objects in the heavens outside our solar system. With this book and your backyard telescope you will be able to identify the brightest and most spectacular star clusters, nebulae, and galaxies — curious objects that inhabit our Milky Way Galaxy — and peer through the cosmos to see distant galaxies far beyond our own.

The objects in this compilation have one thing in common: They are all beautiful when viewed with small telescopes. Beyond that, the objects are distinctly individual. They include 23 galaxies, 19 open star clusters, 18 bright nebulae, 11 globular star clusters, 10 planetary nebulae, 8 double stars, 5 variable stars, 5 dark nebulae, and 1 star cloud.

The objects are listed in order of right ascension, the equivalent of celestial "longitude." The list is confined to the 100 best deep-sky objects. However, because some minor objects lie close by great objects, a total of 110 objects appear in the book. The selection is somewhat skewed toward northern hemisphere sights, although I have included such southern favorites as the Magellanic Clouds, the Coal Sack, and the Eta Carinae Nebula. Four of the objects include other objects in the field: it is difficult to describe star cluster M46 without including the foreground planetary nebula NGC 2438, describe the Andromeda Galaxy without including its satellites M32 and NGC 205, introduce M35 without mentioning its companion NGC 2158, or mention M65 without discussing M66. Three of the objects, the Veil Nebula, the Double Cluster, and the Magellanic Clouds, consist of several parts but are listed as single entries because the individual parts need to be discussed as a whole. Forty-five of the objects belong to that most famous of deep-sky lists, the one compiled by French comet hunter Charles Messier.

In addition to a brief description of each object, this book provides fundamental data, including popular names; designation in the Messier, NGC or other catalog; right ascension and declination coordinates for equinox 2000.0; the constellation the object lies in; V magnitude (except visual magnitudes for planetary nebulae and B magnitudes denoted by subscript B); and size in arcminutes or arcseconds. The data for each object are from Chris Luginbuhl and Brian Skiff's masterly *Observing Handbook and Catalogue of Deep-Sky Objects* (Cambridge University Press, 1990).

May *Beyond the Solar System* introduce you to one hundred friends who will stay with you for a lifetime.

David J. Eicher
Editor-in-Chief, *Deep Sky*
Associate Editor, ASTRONOMY
Waukesha, Wisconsin

Getting to Know the Sky

Astronomy is unlike all other sciences in that its great laboratory is available to everyone on an equal basis. You don't need to be a professional astronomer, in fact, to do astronomy. All you need is a clear sky away from city lights, a pair of binoculars or a small telescope, and a little patience. Indeed, it's much easier to be an instant astronomer than an instant anthropologist. All you need to do is go outside and look skyward.

The challenge in backyard astronomy comes in knowing what to look at. Typically, the first objects people see in a telescope are the Moon and a few planets, usually Saturn, Jupiter, or Mars. (My first observing experience had me looking at Saturn through a friend's 4¼-inch reflector — I was immediately hooked!) But what to look at after the planets? After all, there are a mere eight planets to view and, as fun as solar system observing is, you're going to want to branch out at some point simply for variety's sake.

Beyond the solar system lies the realm of our Galaxy, the Milky Way, and all other galaxies — the basic building blocks of space. If you look at the sky from a numerical viewpoint, virtually everything you can look at with a small telescope lies in the realm of the deep-sky observer. With a 6-inch telescope you can see more than 10,000 deep-sky objects. By "deep-sky" I mean everything beyond the solar system, from the nearest stars to the Galaxy's clouds of gas and dust and groups of stars, to distant galaxies lying millions of light-years away. (A light-year is the distance light travels in one year, approximately equal to six trillion miles.)

If that sounds like an incredible hodgepodge of objects, that's because it is. Here's a short explanation of things that are deep-sky objects and a key to the outstanding examples spread throughout the following pages.

Double and Multiple Stars. These objects are the most plentiful of any for very small telescopes, because the majority of stars that form in the Milky Way are double or multiple systems. Double stars include both genuine articles and frauds. Optical double stars are simply two stars that appear close in the sky because of chance alignment but are far apart in space. Binary stars, however, are true physical systems that orbit a common center of mass and travel throughout the Galaxy locked together by gravity. Multiple stars include stellar systems with three, four, five, or even more suns that coalesced from a single cloud of material. Double and multiple stars are easy to observe and often show beautiful colors ranging from blues, reds, oranges, golds, and tints in between. Outstanding examples in this book include Gamma Andromedae, Zeta Cancri, Porrima, Xi Bootis, and Mizar.

Variable Stars. Due to an array of amazingly different processes, many stars fluctuate in brightness. Extrinsic variables are stars that vary in light output because of an external influence, like being eclipsed by another star. Intrinsic variables vary in brightness due to physical changes in the stars themselves. Numerous classes and subclasses of variable stars exist; for a thorough explanation of the types, see one of the appropriate works listed in the bibliography. Outstanding examples in this book include Mira, Algol, R Coronae Borealis, R Scuti, and SS Cygni.

Open Star Clusters. Stars form from clouds of gas and dust that coalesce until enough mass comes together to turn on nuclear processes. Because of this, stars are born in groups, the products of recycled starstuff, and these groups slowly revolve about the center of the Galaxy until the stars are ultimately dispersed. These open star clusters, loose groups of stars, are visible throughout the disk portion of the Milky Way Galaxy. (Consequently, they're sometimes called galactic star clusters.) Because many are nearby and their stars are relatively bright, open clusters are generally easily visible in small telescopes. Outstanding examples in this book include M35, the Owl Cluster, IC 2602, the Pleiades, the Double Cluster, and the Jewel Box Cluster.

Globular Star Clusters. These curious objects are enormous spheres of old stars floating in the Galaxy's halo. Rather than forming in the galactic disk (like most of the material in the Milky Way), globular clusters formed from leftover clumps of gas and dust. They lie at great distances from us and the galactic center and have enormous, elliptical orbits that carry them plunging down through the Galaxy and out again. Although they contains hundreds of thousands or even millions of stars — as opposed to a few hundred in an open cluster — the individual stars in globulars are difficult to see because of their distances from us. Some resolution into stars is possible with small telescopes, but many globulars appear as fuzzy disks of light. Outstanding examples in this book include 47 Tucanae, Omega Centauri, the Hercules Cluster, M22, M4, and M71.

Planetary Nebulae. An entire class of fuzzy, softly glowing objects, nebulae have a wide range of appearances through telescopes of differing sizes. Planetary nebulae were named so because of their disk-like shapes that make them resemble planets. In actuality, they are spheres of ionized gas slowly expanding from dying stars. In their last gasps of normal life, the stars attempted to burn heavier and heavier material until they gently puffed off shells of gas, reseeding the nearby parts of the Galaxy with material from which to make new stars. Outstanding examples in this book include NGC 1360, the Eskimo Nebula, NGC 6781, the Dumbbell Nebula, and the Ring Nebula.

Bright and Dark Nebulae. This category contains three primary types of nebulae. Emission nebulae are gently glowing gas clouds that, unlike planetaries, are stellar birthplaces in which the gas luminesces by energy from hot young stars. Reflection nebulae do not shine by luminescence, but rather by reflected light from nearby stars. Consequently, they are dim unless they lie closeby. Dark nebulae are opaque dust clouds that block light from stars and gas that lie behind them. Outstanding examples in this book include the Orion Nebula, the Lagoon Nebula, the Omega Nebula, the Rosette Nebula, and the Crab Nebula (bright nebulae); and the S Nebula and Barnard 92 (dark nebulae).

Galaxies. Thus far all of the objects we've looked at have been members of our own Galaxy. Although globulars, planetaries, emission nebulae, open clusters, and double stars all exist in other galaxies, the galaxies are so distant that such details are not readily visible in backyard telescopes. But we can observe galaxies as large structures — huge islands of stars that from the outside appear like balls of light, smooth disks, barred spirals, or mottled, dusty edge-on streaks of light. Outstanding examples in this book include the Andromeda Galaxy, the Whirlpool Galaxy, M81, M101, Centaurus A, the Sombrero Galaxy, and M33.

A Personal Exploration of the Universe. To begin your exploration of the best deep-sky wonders, you'll want to practice using your binoculars or small telescope until you're comfortable with it. Obtain a reliable star atlas that shows the brightest deep-sky objects (see bibliography). The best ones are *Sky Atlas 2000.0* by Wil Tirion, *The Observer's Sky Atlas* by Erich Karkoschka, *Uranometria 2000.0* by Wil Tirion, George Lovi, and Barry Rappaport, and *Norton's 2000.0* by Ian Ridpath. To begin your tour of the greatest 100, use this book in conjunction with one of the atlases to locate the objects, study them, and keep an observing log of what you've seen. I hope that by observing the objects listed in this book you'll enjoy your introduction to astronomy's great laboratory in the sky.

Photo courtesy Harvard College Observatory

1. 47 Tucanae (NGC 104)
Globular cluster
Right Ascension and Declination: 0h24.1m -72°05'
(Equinox 2000.0)
Constellation: Tucana
Magnitude: 4.0
Size: 30.9'

The globular cluster 47 Tucanae appears as a bloated, 4th-magnitude "star" to the naked eye. This object's proximity to the Small Magellanic Cloud — they are separated by a mere 1° — makes the field surrounding them appear like a subtly glowing painting to those far enough south to see it. Although 47 Tucanae is smaller and dimmer than the cluster Omega Centauri, a telescopic view of 47 Tucanae shows it is the most impressive cluster in the sky. The so-called giant branch stars in 47 Tucanae are significantly brighter than those in most globular clusters, so rather than showing dozens or hundreds of stars resolved, small telescopes resolve literally thousands of stars across the face of 47 Tucanae. A 6-inch telescope at 100x shows pinpoint specks across a 25-foot-diameter blazing ball of light. The center of the cluster appears as a milky disk of light because innumerable stars "pile up" on each other. Most deep-sky observers believe 47 Tucanae is the most beautiful globular cluster in the sky.

2. The Andromeda Galaxy
(M31, NGC 224)
Spiral galaxy
R.A. and Dec.: 0h42.7m +41°16'
Con.: Andromeda
Mag.: 3.5
Size: 180' by 63'

NGC 205
Elliptical galaxy
R.A. and Dec.: 0h40.3m +41°41'
Con.: Andromeda
Mag.: 8.0
Size: 17' by 9.8'

M32 (NGC 221)
Elliptical galaxy
R.A. and Dec.: 0h42.7m +40°52'
Con.: Andromeda
Mag.: 8.2
Size: 7.6' by 5.8'

The Andromeda Galaxy is the Milky Way's sister spiral in the Local Group of galaxies. Because it is the nearest spiral galaxy to our own, it is visible to the naked eye as an elongated streak about 2° long. On moonless nights, the Andromeda Galaxy unleashes a wealth of telescopic detail. Binoculars or a 2-inch telescope reveal a bright, oval glow measuring 2.5° by 1° in extent and cast in an eerie, greenish-white hue. The central part of the galaxy is the most intense, with an almost uniformly bright oval disk constituting the galaxy's core. Surrounding this bright glow is an oval ring of much lower surface brightness.

A 4-inch scope at low power easily shows the galaxy's two primary satellites, the small ellipticals M32 and NGC 205. M32 is visible on the Andromeda Galaxy's southern edge, while NGC 205 — larger, fainter, and more elongated than M32 — lies northwest of the main galaxy's center. On nights of exceptional clarity, a 4-inch scope will show two broad dust lanes on the northern side of the Andromeda Galaxy's central bulge. Also visible is the small star cloud NGC 206, a small patch of light 30' west-southwest of M32. If this dense cloud of stars is visible, try moving to low power and getting a glimpse of the galaxy's lovely spiral pattern.

Photo by Tony Hallas and Daphne Mount

3. NGC 253
Spiral galaxy
R.A. and Dec.: 0h47.6m -25°18'
Con.: Sculptor
Mag.: 7.1
Size: 25' by 7.4'

NGC 253 is the southern sky's answer to the Andromeda Galaxy. Although smaller and dimmer than M31, NGC 253 is an impressive sight and is easily visible in binoculars or small telescopes. A perfect spiral, NGC 253 is inclined so that it appears to be a tipped dinner plate. Located in Sculptor at a declination of -25°, NGC 253 is not easily visible for northern hemisphere viewers, so the galaxy's impressive appearance — even relatively low in the sky — testifies to its great brightness.

A 4-inch telescope shows a slender streak of milky white light with two bright field stars immediately south of the galaxy's center. On dark nights the galaxy's mottled texture is visible, especially with averted vision. An 8- or 10-inch telescope at high power shows the mottled appearance clearly, suggesting NGC 253's clumpy spiral arm pattern.

Photo by David Healy

4. The Magellanic Clouds

The Small Magellanic Cloud
Spiral galaxy
R.A. and Dec.: 0h52.7m -72°50'
Con.: Tucana
Mag.: 2.3
Size: 280' by 160'

The Large Magellanic Cloud
Spiral galaxy
R.A. and Dec.: 5h23.6m -69°45'
Con.: Dorado
Mag.: 0.1
Size: 650' by 550'

The Small and Large Magellanic clouds are the two satellite galaxies that orbit our own Milky Way. Because they are so close, the Magellanic clouds are prominent features of the sky. Each lies deep in the southern hemisphere sky, so they are not visible from North America.

The Magellanic clouds are huge and exceedingly bright. Tucana's SMC covers nearly 5° by 3° and shines with a total magnitude of 2.3. The Large Magellanic Cloud, located in Dorado, spans almost 11° by 9° and shines with the light of a 0.1-magnitude star. Because each of the clouds is so large and is an individual galaxy in itself, it is filled with a variety of its own deep-sky objects. Besides lying adjacent to globular clusters NGC 362 and 47 Tucanae (both Milky Way objects), the pear-shaped SMC has clumps and knots of material that are visible with large binoculars or small telescopes. The bar-shaped LMC is magnificently more detailed: It holds the impressive emission nebula dubbed NGC 2070, the Tarantula Nebula, and a slew of fainter objects. Each of the clouds makes a fascinating region to scan with a small scope on a dark night.

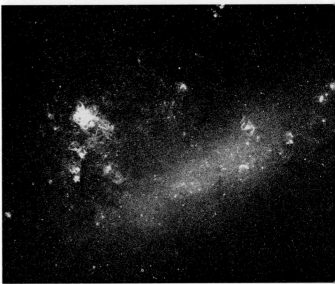
Photo courtesy Kitt Peak National Observatory

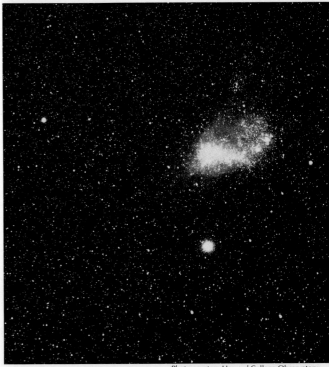
Photo courtesy Harvard College Observatory

Photo by Ronald Royer

5. NGC 362
Globular cluster
R.A. and Dec.: 1h03.2m -70°51'
Con.: Tucana
Mag.: 6.6
Size: 12.9'

NGC 362 is one of the southern hemisphere's premier globular star clusters. The cluster's proximity to the Small Magellanic Cloud — they are separated by about 1° — provides binoculars or small telescope users with plenty to see in this area.

A 3-inch telescope at high power shows NGC 362 as a milky white disk some 8' across. A 6-inch telescope nicely resolves the brightest cluster members, showing a 10'-diameter glow peppered with several dozen faint stars, mostly on the cluster's edges. NGC 362 is typically overshadowed by the closeness of the spectacular cluster 47 Tucanae; however, each is unique and warrants close examination.

Photo courtesy Cerro Tololo Interamerican Observatory

6. The Owl Cluster (NGC 457)
Open cluster
R.A. and Dec.: 1h19.1m +58°20'
Con.: Cassiopeia
Mag.: 6.4
Size: 12'

A loose, bright open star cluster, NGC 457 lies in the W-shaped constellation Cassiopeia. One night in 1977 I tracked down this object by zeroing in on the star Phi Cassiopeiae and began to study the cluster's form. Suddenly it struck me that the stars clearly formed the shape of an owl. Two bright stars — orangy, 5th-magnitude Phi Cas and another somewhat fainter star — form eyes, while a trail of fainter stars represents what appear to be wings, a head, body, and feet. Ever since that night I've called this object the Owl Cluster.

Because it measures 12' across and lies in a rich star field, use a low- to medium power eyepiece when you observe NGC 457. A 6-inch scope at 50x shows nearly 75 stars in this sparkling group plus a uniformly spread background glow from the starry Cassiopeia Milky Way. After seeing how pretty and easy to find this area is, the NGC 457 field may become one of your favorite northern sky sights.

Photo by Lee C. Coombs

7. M33 (NGC 598)
Spiral galaxy
R.A. and Dec.: 1h33.9m +30°39'
Con.: Triangulum
Mag.: 5.7
Size: 62' by 39'

Sometimes called the Pinwheel Galaxy, the challenging galaxy M33 can be found in the tiny constellation Triangulum. Although the galaxy's total brightness is great, its light is spread over an enormous area, which makes its surface brightness — the brightness per unit of area — low. Thus, some observers report spotting M33 at the limit of their naked-eye vision on dark nights. Others, however, find the galaxy hard to spot and disappointing even in relatively large backyard telescopes.

Because it is a low-contrast object, wait until a moonless night to search for M33. Binoculars or a wide-field telescope are the best tools, since the galaxy measures 1° by 0.6° in extent. Slowly sweep over M33's position and you will see a large, faint core with an almost starlike center. The galaxy's huge, ghostly spiral arms may be visible, almost resembling a soft, glowing blanket tucked around the galaxy's central hub.

Photo by Kim Zussman

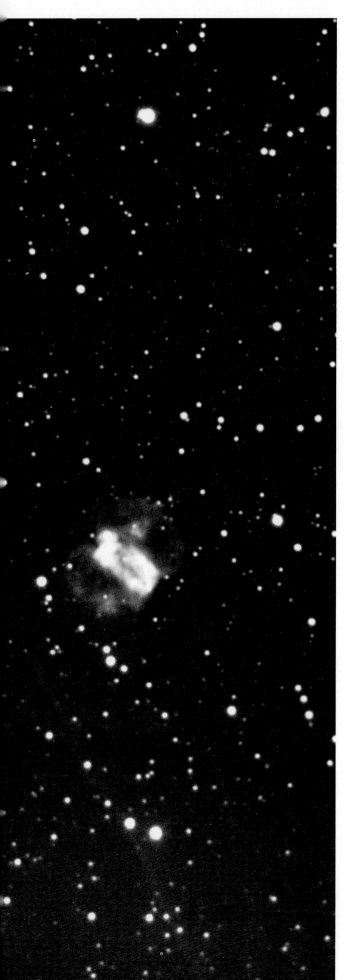

Photo by Kim Zussman

8. The Little Dumbbell Nebula
(M76, NGC 650-1)
Planetary nebula
R.A. and Dec.: 1h42.4m +51°34'
Con.: Perseus
Mag.: 10.1
Size: 2.7' by 1.8'

The diminutive planetary nebula M76 is a challenging object for small telescopes. Consisting of two dim lobes on either side of a 16th-magnitude central star, M76 acquired the nickname Little Dumbbell Nebula because of its resemblance to its bigger, brighter cousin M27, in Vulpecula. Successfully observing the Little Dumbbell Nebula with a small telescope requires a dark night and a moderately high-power eyepiece. On faint objects like M76 a nebular filter helps significantly by boosting contrast between the object and sky background. A 6-inch telescope at high power shows M76 as a tiny pair of round glows in contact, forming a luminous bar. On extremely dark nights you may see a faint glow surrounding this bar-shaped patch of nebulosity. In larger telescopes, M76 is more striking and complex. A 17.5-inch scope shows the little planetary as a bright bar of material with two faint semicircular rings of nebulosity on each side. These outer rings have a low surface brightness and normally require averted vision to be seen.

9. Gamma Andromedae
Double star
R.A. and Dec.: 2h03.9m +42°20'
Con.: Andromeda
Mags.: 2.3, 5.5
Sep.: 9.8"

For small telescope users, the star Gamma Andromedae is well worth viewing on autumn evenings when the air is cool and steady. Composed of a magnitude 2.3 star with a strong golden yellow glow (some say tinged with orange) and a magnitude 5.5 greenish-blue secondary, Gamma Andromedae offers a striking color contrast in any scope. The stars are separated by a generous 9.8" in position angle 63° (1967), making the gap between them easily visible in any instrument at medium powers. This inky black gap represents over 74 billion miles of space between the two stars. Also known as Almach, Gamma lies in the east-central part of Andromeda about 5° northeast of the bright open star cluster NGC 752. Curiously, the primary star is itself a double, but it is virtually impossible to split because of its separation (0.3"). Additionally, the brightest star in the primary system is a spectroscopic double star, making Gamma Andromedae a system with four stars.

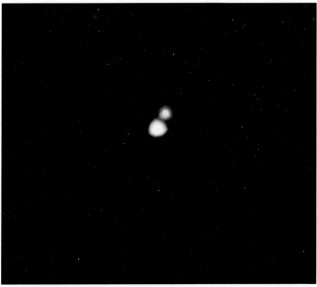

Photo by Ron Miller

10. The Double Cluster

NGC 869
Open cluster
R.A. and Dec.: 2h19.0m +56°09'
Con.: Perseus
Mag.: 3.5
Size: 18'

NGC 884
Open cluster
R.A. and Dec.: 2h22.4m +55°07'
Con.: Perseus
Mag.: 3.6
Size: 18'

Lying between the bright stars of Perseus and those of Cassiopeia, the Double Cluster is a great galactic coincidence. As we see them, these two open clusters cover the same area of sky, shine with nearly the same brightness, and are located within 1° of each other. They are, however, physically unrelated: the clusters are separated by 400 light-years of space. (NGC 884 is 7,500 light-years distant, while NGC 869 is a mere 7,100 light-years away.) The fact that these clusters are not physically associated should not prevent you from enjoying viewing them. The best method is to use binoculars or a small telescope with the lowest possible magnification. In this way you'll capture the greatest field of view possible and see all of the stars in both clusters at once. When you center your scope on the cluster you'll see a field alive with stars. NGC 869 contains about 400 stars brighter than 12th magnitude, while NGC 884 holds 300. The brightest stars in the clusters shine at about magnitude 6.4, providing the clusters with enough luminosity to be visible with the naked eye. In telescopes, the pretty star colors will be striking. Many bright blue and blue-white stars are visible, and a number of the fainter stars in each cluster are rose colored.

Photo by Martin C. Germano

11. Mira (Omicron Ceti)
Long period variable star
R.A. and Dec.: 2h19.3m -2°59'
Con.: Cetus
Mag. range: 2.0-10.1
Period: 332 days

This famous variable star was discovered in 1596 by the Dutch astronomer David Fabricius. Ever since, it has been observed by those curious about the periodic fluctuations of star light. It has even been nicknamed "The Wonderful" because of its dramatic increases and decreases in brightness.

Mira is a red giant star, the brightest of the long-period pulsating variable stars. Typically the star bounces back and forth between magnitude 10, where it stays at minimum and magnitude 3 or 4 at maximum. The star occasionally flares to 2nd magnitude and, in 1779, astronomers watched Mira explode to 1st magnitude in brightness and rival the star Aldebaran.

What causes these stellar flare ups? Mira is a star that physically pulses in size and brightness. At minimum it is approximately 400 times the diameter of the Sun and one of the coolest stars known. At maximum, however, the star swells to more than 500 times the diameter of the Sun, its temperature rises, and the star's color shifts slightly away from red into orange. This pulsation is most likely caused by the old star's change from straight hydrogen burning into the early stages of helium burning. The theory of pulsation, however, is not thoroughly understood.

Photo by Robert Provin.

Photo by Robert Provin and Brad Wallis

12. Algol (Beta Persei)
Eclipsing binary star
R.A. and Dec.: 3h08.2m +40°57'
Con.: Perseus
Mag. range: 2.1-3.4
Period: 2.9 days

Unlike Mira, a physically changing star, the bright variable star Algol is an eclipsing binary. In other words, the periodic variations in light output are caused by a massive, dark object blocking the light from the bright component. Nicknamed the Demon Star, Algol is thought to have been observed as a variable for several centuries. The first definitive notes on the matter were made by the Italian astronomer Geminiano Montanari in 1667. Two hundred and twelve years later the German astronomer H. C. Vogel showed that Algol's variability is caused by a periodic eclipse. Because of the short time scale involved, Algol is one of the most easily observed variables in the sky for small telescope astronomers. The star normally shines at magnitude 2.1, but every 2.9 days it fades to magnitude 3.4 and slowly returns to normal brightness over an eclipse duration of 10 hours. Observe this star each night over a period of several days, sketch the field, and note the brightness of the surrounding stars. You'll see that Algol's magnitude does indeed dramatically change.

Photo by Martin C. Germano

13. NGC 1360
Planetary nebula
R.A. and Dec.: 3h33.2m -26°52'
Con.: Fornax
Mag.: 9.4
Size: 9' by 5'

Because NGC 1360 lies in an obscure part of an obscure constellation — and is low in the sky for Northern Hemisphere viewers — it has received little attention. Too bad. In fact, it is relatively easy to find.

Start by locating the 2nd-magnitude star Beta Ceti, the brightest star in this part of the sky. From Beta, move approximately 15° east and 3° south to the 4th-magnitude star Upsilon Ceti, then another 12° east and 8° south to the 4th-magnitude double star Alpha Fornacis. From there, aim your finderscope 4° north and 5° east, and you'll be centered on NGC 1360 in a field that contains the bright variable star RZ Fornacis.

As planetaries go, NGC 1360 is both big and bright — as bright as the Ring Nebula but four times larger. The morphology of NGC 1360 is certainly peculiar, because even a small telescope shows an object like a glowing lozenge with a speckled, mottled appearance and a peppering of faint stars involved with the nebulosity. The most striking characteristic of this nebula is its strong blue color, an attribute that makes it easy to pick out in a small finderscope.

14. The Pleiades (M45, Melotte 22)
Open cluster
R.A. and Dec.: 3h47.0m +24°07'
Con.: Taurus
Mag.: 1.2
Size: 120'

Large, bright, and possessing a distinctive shape, the Pleiades is universally revered as the greatest open cluster in the northern sky. The group's nickname, The Seven Sisters, grew out of mythology and was applied to the cluster because of its seven brightest stars, which are arranged in a dipper shape. The seven are (with magnitudes), Alcyone (Eta Tauri, mag. 2.9), Atlas (mag. 3.6), Electra (mag. 3.7), Maia (mag. 3.9), Merope (mag. 4.2), Taygeta (mag. 4.3), and Pleione (mag. 5.1). Besides these bright stars, 500 or more cluster members are visible in backyard telescopes, over a diameter of 120 — an area four times the width of the Full Moon.

At magnitude 1.2, the Pleiades is easily visible to the naked eye. During autumn and winter months it's usually what people see when they glance at the sky and say, "Hey — what's that fuzzy patch of light?" The best instruments for studying the Pleiades are a good pair of binoculars or a telescope with a low-power, wide field of view. With a scope, slowly scan the bright stars and note the strong blue-white colors and the many pairs and strings of stars throughout the group. On especially dark nights, use a blue-transmissive nebular filter to observe a patch of reflection nebulosity near the star Merope. This nebulosity, designated NGC 1435, is extremely faint but within the reach of backyard instruments. Even if you fail to spot the elusive nebulosity, the colorful star patterns in the Pleiades will keep you attached to the eyepiece for long periods as you scan the area.

Photo by Rick Dilsizian

15. The California Nebula (NGC 1499)
Emission nebula
R.A. and Dec.: 4h03.4m +36°25'
Con.: Perseus
Mag.: —
Size: 160' by 40'

Large and exceedingly faint, the California Nebula in Perseus is a long time favorite for astrophotographers because its ruddy hue shows up well on color film. The low surface brightness object, however, is a viable target for deep-sky observers — under the right conditions.

To observe the California Nebula, check a star atlas and locate the correct position on the sky. First, find the bright star Xi Persei, the hot sun that excites the California Nebula into luminescing. The California is immediately north of the star and within the same low-power field of view. Because the nebula measures more than five Moon diameters in length, use the lowest possible magnification. If you have a red-transmissive nebula filter, it will help to boost contrast between the nebula and the sky background.

On a moonless night, you'll see the California as a smooth, oval glow with minor variations in surface brightness across the nebula. With telescopes of 10-inches or more aperture you may spot with averted vision some of the filamentary structure in the California that is readily visible on long-exposure photographs.

Photo by Alfred Lilge

Photo by J.A. Farrell

16. The Hyades (Melotte 25)
Open cluster
R.A. and Dec.: 4h26.9m +15°52'
Con.: Taurus
Mag.: 0.5
Size: 2100'

As you stroll under the wintertime stars you'll see a bright, V-shaped group of stars that forms the central part of Taurus. This entire group — with the exception of 1st-magnitude Aldebaran — is made up of stars held together by gravity in a loose cluster called the Hyades. At a distance of 130 light-years, the Hyades cluster is the second closest open cluster in our sky. (The closest is the Ursa Major Moving Group, a collection of stars so close that it appears scattered across much of the sky.) Consequently, many of the Hyades stars are exceptionally bright. The brightest are $Theta^2$ Tauri (mag. 3.3), Epsilon Tauri (mag. 3.5), Gamma Tauri (mag. 3.7), Delta Tauri (mag. 3.8), $Theta^1$ Tauri (mag. 3.9), Kappa Tauri (mag. 4.1), Upsilon Tauri (mag. 4.2), 90 Tauri (mag. 4.2), 68 Tauri (mag. 4.2), and 71 Tauri (mag. 4.4).

Although brighter and larger than the nearby Pleiades cluster, the Hyades is not so impressive in telescopes. Rather, the Hyades is so spread out it is a perfect hunting ground for binocular observers. Scan the cluster and see how many star colors, patterns, and pairs you can pick up. The several hundred stars in the Hyades offer a great variety of binocular sights.

Photo by Neyle Sollee

17. The Crab Nebula (M1, NGC 1952)
Supernova remnant
R.A. and Dec.: 5h34.5m +21°01'
Con.: Taurus
Mag.: —
Size: 6' by 4'

The Crab Nebula is the object that started much of the current interest in deep-sky observing. In September 1758, French comet hunter Charles Messier found a nebula "in the horn of Taurus" that appeared like a comet yet remained fixed in location. So he and other comet hunters wouldn't confuse these nebulae with comets — which move across the sky — Messier started a catalog of nebulae, which became the standard "hit list" of objects for deep-sky observers. The object in Taurus became Messier 1 and was subsequently dubbed the Crab Nebula after its tiny filaments visible in photographs.

The Crab Nebula is a special type of a bright nebula. Most are made up from dust glowing by reflected light or warm gas coalescing into infant suns, but the Crab is a supernova remnant. In 1054 A.D. Chinese astronomers observed a "new star" in Taurus, and the Crab is the exploded remains of this dead sun as we see it some 940 years later. This is such an object of change that large telescope photos taken at the beginning of the 20th century and those made today are different enough to show the gas expanding in space.

The Crab is easy to find and relatively bright, making it a popular target for small scopes. Start by locating Zeta Tauri, the star marking the southern tip of the bull's horn. From there, move 1.5° northwest and the Crab will come into view. If you're using a small telescope, you'll see a small glow about 2' by 4' across and glowing at about 9th magnitude. Larger telescopes begin to show detail in the Crab: a 10-inch telescope reveals uneven mottling across the surface of the nebula, several bright stars involved with the nebulosity, and a hint of the twisted filaments that glow subtly against the inky black sky.

18. The Orion Nebula (M42, NGC 1976)
Emission nebula
R.A. and Dec.: 5h35.4m -5°23'
Con.: Orion
Mag.: —
Size: 90' by 60'

Offering great detail visible in backyard telescopes, M42, the bright, fuzzy patch of light located in Orion's Sword, is the most beloved of all emission nebulae. Easily visible to the naked eye and unquestionably observed since prehistoric times, the Orion Nebula is the central showpiece of the winter sky. Its great brightness and huge dimensions — in area it equals 7.5 Full Moons — make it unequaled by other nebulae in the northern sky.

The Orion Nebula is a large star-forming region located relatively close by in our Galaxy. It is composed of warm, luminescing gas, recycled from dead stars, that is slowly coming together and producing new stars by gravitational collapse. Many of the young suns forming within the Orion Nebula are visible in small telescopes.

With a 4-inch reflector, for example, you can observe the multiple star that lies near the center of M42. Called the Trapezium, Theta[1] Orionis is a system made up of four bright stars glowing between magnitudes 5.1 and 7.9 and separated by 8.8" and 19.3". That central group is so distinctive in small scopes that some observers nearly don't notice Theta[2] Orionis, a wide double star lying several arcminutes to the northwest. A smattering of fainter stars lies scattered throughout the region, most of which are stars born out of the Orion Nebula. The nebula itself appears as a peculiarly shaped halo of greenish light enveloping the field and extending north and south for some distance.

Photo by Steve Quirk

Photo by Preston S. Justis

19. The Tarantula Nebula (NGC 2070)
Emission nebula
R.A. and Dec.: 5h38.7m -69°06'
Con.: Dorado
Mag.: —
Size: 40' by 25'

A great counterpart to the Orion Nebula in the southern sky is the Tarantula Nebula, another large, bright cloud of emission nebulosity. The Tarantula would be much larger and brighter if it were located in our Galaxy rather than inside the Large Magellanic Cloud, at the great distance of 170,000 light-years. Indeed, the Tarantula is the largest emission nebula currently known in the universe. (See entry number 4 for a discussion of the Magellanic Clouds.)

Observers in the Southern Hemisphere can find this object in the constellation Dorado by first finding Canopus. From there, move south-southwest 12° to a trapezoid of bright stars that includes Beta, Delta, and G Doradus. From Delta, drop 4° south into the fuzzy glow of the Large Magellanic Cloud and you'll come upon the Tarantula Nebula. Despite its great distance, the Tarantula is a spectacular object in the eyepiece because it is so large and complex. (If the Tarantula were as close as the Orion Nebula, it would cover 30° of sky!) As it is, the naked-eye object fills a telescopic field with a bright splash of glowing light surrounding a group of faint stars scattered across the field. Great curving loops of nebulosity are visible in small telescopes, and large instruments show a complex field packed with intricate details.

20. Horsehead Nebula (Barnard 33)
Dark nebula
R.A. and Dec.: 5h40.9m -2°28'
Con.: Orion
Mag.: —
Size: 6' by 4'

The Horsehead Nebula in Orion is a grand example of an object that has become increasingly popular with backyard observers — dark nebulae. (This popularity is due to the widespread availability of large backyard telescopes that make observing these challenging objects feasible.) These clouds of fine, sooty dust particles presumably lie scattered in thick blankets all across the Galaxy, but are only easily visible when they lie nearby and are lit from behind by bright objects. This is exactly the case with the Horsehead. A faintly glowing emission nebula designated IC 434, which extends southward from near the star Zeta Orionis and illuminates the Horsehead enough for us to see it. (The Horsehead Nebula acquired its nickname because it resembles the profile of a horse's head.)

Edward C. Pickering discovered the Horsehead Nebula on photographs made in 1889. That it was discovered photographically does not mean it is invisible to the eye, however. Observers with a wide range of instruments — from 4-inch refractors to 30-inch reflectors — have reported seeing the Horsehead, although it is small and offers only low contrast to the visual observer. Outfitting your telescope with the proper nebula filter and observing when the nebula is near the meridian will help bring this difficult object closer to visual reach.

Photo by Tony Hallas and Daphne Mount

21. M37 (NGC 2099)
Open cluster
R.A. and Dec.: 5h52.4m +32°33'
Con.: Auriga
Mag.: 5.6
Size: 20'

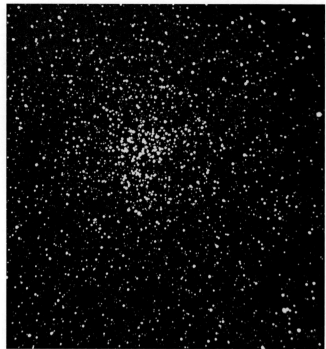
Photo by Martin C. Germano

M37 is the prettiest open star cluster in a constellation loaded with pretty open star clusters. Discovered by Charles Messier in 1764, this object has since become a favorite with amateur astronomers.

To locate M37, start by finding the 2nd-magnitude star Beta Tauri. Move your finderscope 3.5° northeast until you spot the 5th-magnitude double star 26 Aurigae. Now move another 3.5° northeast along the same line, and you'll be sitting right on M37. This cluster is one of a few that appears almost unbelievably rich in small scopes, in that it contains hundreds of stars packed into a relatively small area. The 19th-century observer C. E. Barns called it "a diamond sunburst." His contemporary, William Henry Smyth, recorded that the "whole field [was] strewn as if it were sparkling with gold dust." Altogether, M37 contains 150 stars brighter than 12th magnitude, all of which are visible in small scopes. The cluster's overall shape is somewhat triangular, and a bright, orange star is placed dramatically near the cluster's center. The distance to this group is approximately 4,600 light-years.

22. M35 (NGC 2168)
Open cluster
R.A. and Dec.: 6h08.9m +24°20'
Con.: Gemini
Mag.: 5.1
Size: 28'

NGC 2158
Open cluster
R.A. and Dec.: 6h07.5m +24°06'
Con.: Gemini
Mag.: 8.6
Size: 5'

Photo by Martin C. Germano

Another fine open cluster in the winter sky lies in Gemini. M35 is large and bright enough that it is best observed with binoculars, having been discovered in antiquity by unknown naked-eye observers. The whole group appears to the naked eye as a fuzzy glow somewhat smaller than the Moon but distinctly visible with direct vision under a dark sky.

M35 is conveniently located in the feet of Gemini, in a region of particularly bright stars. To find it, locate the 4th-magnitude stars Eta and Mu Geminorum, and the 5th-magnitude sun 1 Geminorum. M35 lies 2° north and slightly east of 1 Geminorum.

A 4-inch scope shows M35 as a smattering of 7th-, 8th-, and 9th-magnitude stars. Larger telescopes bring out many fainter stars that glow eerily white, yellow, and pale orange. Altogether, 500 stars lie within an area slightly smaller than the diameter of the Full Moon. As a bonus, observers with telescopes 6-inch and larger in aperture will detect the distant open cluster NGC 2158 lying immediately southwest of M35's edge. NGC 2158 is a physically similar object to M35 that lies six times farther away. Resolving NGC 2158 into stars is difficult because the brightest members glow feebly at 16th magnitude. However, most small telescopes show the cluster as a hazy glow about 3' across.

23. The Rosette Nebula (NGC 2237-9)
Emission nebula
R.A. and Dec.: 6h32.3m +4°38'
Con.: Monoceros
Mag.: —
Size: 90' by 90'

Although it is unimpressive to the naked eye, the dim constellation Monoceros is home to the wreath-shaped emission nebula NGC 2237-9, nicknamed the Rosette Nebula. Spectacularly detailed in amateur photographs, the Rosette is a challenging object to see visually because of low surface brightness. But on dark nights, large binoculars or wide-field scopes reveal a smooth, milky white glow surrounding the bright star cluster NGC 2244. (The star cluster is physically associated with the nebula, being composed of the suns coalescing out of the envelope of gas that permeates them.)

To locate the Rosette Nebula, start at the bright star Betelgeuse in Orion and move 9° east to the 5th-magnitude star 13 Monocerotis. Next, move south 3°, and the star cluster will come into view in your binoculars or finderscope. With the Rosette, getting there is the easy part — seeing the nebulosity is the hard part. One simple technique that helps to spot the faint nebula is slowly sweeping your telescope back and forth. It's much easier initially to detect faint light if the object in question is slowly moving. Additionally, if you have a red transmissive nebula filter, use it. The filter will allow the light from the nebula to pass through the eyepiece while blocking stray skylight. Finally, use averted vision — glancing off to the side of the telescope's field of view. This allows the light from faint objects like the Rosette to fall on the most sensitive detectors in your eye. If you use these techniques, a dark night will help reveal the Rosette as a faint circular glow 1° across shimmering with a ghostly green hue.

Photo by Tony Hallas and Daphne Mount

24. Hubble's Variable Nebula
(NGC 2261)
Emission and reflection nebula
R.A. and Dec.: 6h39.1m +8°43'
Con.: Monoceros
Mag.: —
Size: variable

Most deep-sky objects are static objects as viewed from Earth. Actually, they change dynamically, yet their great sizes and immense distances render our views of them mere still frames in an ongoing cosmic movie. But there are rare exceptions, one of which is a small reflection nebula in Monoceros. Like all reflection nebulae, Hubble's Variable Nebula is a cloud of dark dust made visible by a hot star that illuminates it from within. In the case of Hubble's Variable, however, the illuminating star is a variable star, R Monocerotis. This star irregularly bounces between magnitudes 10 and 12. When the star dims, the nebula shrinks. When the star brightens, the nebula grows. The effect is dramatic enough to be visible in backyard telescopes.

Hubble's Variable Nebula lies just a few degrees from the Rosette Nebula. Start at the 5th-magnitude star 13 Monocerotis and move 3° northeast. That will place you in the correct low-power field to find the nebula. Once there, use a moderately high-power eyepiece (about 100x) to yield enough image scale to see detail in the nebula. If you have a blue transmissive nebula filter, use it. The nebula will appear as a dim, almost triangular-shaped wedge of fuzzy light. Because it changes, this nebula is a good one to sketch periodically. You may find over months and years that its dimensions and brightness vary to a startling degree.

Photo by Paul Roques

25. The Cone Nebula (NGC 2264)
Emission and dark nebula
R.A. and Dec.: 6h41.0m +9°54'
Con.: Monoceros
Mag.: —
Size: 10' by 7'

Two degrees north of Hubble's Variable Nebula lies a huge complex of young stars and emission nebulosity roughly comparable to the Rosette Nebula. This time the central star cluster — the young suns forming from the nebulosity — faintly resembles the outline of a tree (this object is often called the Christmas Tree Cluster). The main target of interest here is the faint shell of nebulosity that surrounds the cluster. A large, lacy background of warm, luminescing gas is involved with the cluster and extends several degrees in each direction. A particular part of the nebula — 1° due south of the star cluster — is the place to center your eyepiece.

At this spot you'll see a small, cylindrical notch in the faint outline of the nebula. This is the Cone Nebula itself, a dark nebula measuring 10' by 7'. This object is challenging for small telescopes unless the night is dark and transparent, but once found it will become one of your favorite dark nebulae in the winter sky.

Photo by Martin C. Germano

26. Sirius (Alpha Canis Majoris)
Double star
R.A. and Dec.: 6h45.1m -16°43'
Con.: Canis Major
Mags.: -1.5, 8.5
Sep.: 5.5"

Sirius is a star of extremes. At magnitude -1.5, it is the brightest star in the sky, the crown jewel of Canis Major and indeed the dominant gem of a richly studded winter evening sky. Its white glare is so strong that most deep-sky observers avoid it completely, fearing that staring at such a bright object will temporarily ruin their dark adaptation — their ability to see details in faint galaxies and nebulae. Sirius is not merely a super-bright star, however. It is a double star, one with components of extremely different brightnesses.

The second star in the Sirius system, called Sirius B, is a tiny white dwarf. Whereas the main star shines with the brilliance of 23 Suns, the companion has a feeble light output equaling only 1/500th that of the Sun. Therefore, although the two stars are separated by a generous 5.5", the overpowering glare from Sirius A makes catching a glimpse of the magnitude 8.5 secondary frightfully difficult.

The best method is to employ high magnification on a night of steady seeing and scan the area of the secondary while trying to keep the primary star just out of the field. Another technique is to lay an occulting bar in or on the eyepiece to block the glare from the primary. In any case, as time passes, the elusive secondary will become easier to see. The stars orbit each other about every 50 years, and the separation will begin to widen about the turn of the century.

Photo by Jim Barclay

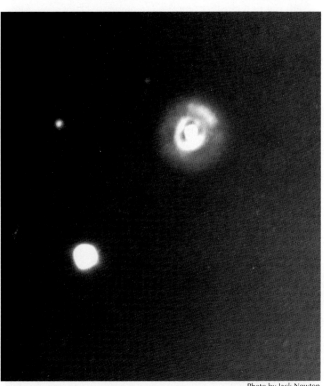

Photo by Jack Newton

27. M41 (NGC 2287)
Open cluster
R.A. and Dec.: 6h47.0m -20°44'
Con.: Canis Major
Mag.: 4.5
Size: 38'

Aim your telescope or binoculars 4.5° south of Sirius and you'll discover the sparkling open cluster M41. Composed of two dozen bright stars and 100 faint ones, this loose group of stars is larger than the Full Moon and, at magnitude 4.5, is visible to the naked eye under a reasonably dark sky.

In a telescope M41 is a splendid sight, consisting of a handful of bright yellow and orange stars as well as a myriad of blue and white ones. The view overall is of a rich Milky Way star field with the cluster hovering in the foreground, the cluster stars shining between magnitudes 7 and 13. The stars in M41 are intrinsically bright: if the light output from all of M41's stars could be combined, it would equal that of 1500 suns.

Photo by Martin C. Germano

28. The Eskimo Nebula (NGC 2392)
Planetary nebula
R.A. and Dec.: 7h29.2m +21°55'
Con.: Gemini
Mag.: 9.2
Size: 47" by 43"

Most planetary nebulae appear as bright, blue-green disks that are slightly larger than stars. Some are exceptions, however, and the Eskimo Nebula in Gemini is one of these. This nebula consists of a bright disk measuring 40" across. Surrounding this disk is a faint, tenuous halo of broken light that in photographs appears like a broken ring. In observatory photos, the overall effect is that of the face — the central star representing the nose — surrounded by a parka or fur-lined hood. Thus, astronomers long ago coined the term Eskimo Nebula to describe NGC 2392.

To locate this object, aim your telescope at the 3rd-magnitude double star Delta Geminorum. First, move 2° east-southeast to a bright clump of stars that contains the 5th-magnitude double star 63 Geminorum. Now swing your telescope's field 1° southeast and you'll be centered on the correct field.

Like most planetary nebulae, NGC 2392 has a high surface brightness and bears magnification well. A 6-inch scope at high power shows a bright blue disk surrounding a 10th-magnitude central star. Visually, it's extremely difficult if not impossible to spot the outer "parka" of nebulosity. Large telescopes with specialized nebula filters may show some detail within the disk on nights of extraordinary seeing.

29. NGC 2403
Spiral galaxy
R.A. and Dec.: 7h36.9m +65°36'
Con.: Camelopardalis
Mag.: 8.4
Size: 18' by 11'

Large spiral galaxies are fun to observe. They provide deep-sky viewers with the chance to see spiral arms, clumps of starlight from stellar associations, dust patches, and the starlike nucleus of a distant island universe. On spring nights, try spotting NGC 2403, one of the nearest spiral galaxies beyond our own Local Group.

The easiest way to find NGC 2403 — a galaxy isolated in a stark region of the sky — is to first aim your scope at Merek (Beta Ursae Majoris), the star that marks the bottom front edge of the Big Dipper. Move 5° north to Dubhe (Alpha Ursae Majoris), then make a right turn and head west 10° to the 4th-magnitude double star 23 Ursae Majoris. Now move just short of an equal distance west and slightly south to a somewhat brighter double star, Omicron Ursae Majoris. Move an additional 10° west and 5° north and you'll be pointing at the NGC 2403 field.

NGC 2403 is a Sc-type galaxy tipped slightly from face-on, so a great amount of detail is visible in the galaxy's spiral arms. A good 6-inch instrument will faintly show the arms on a dark night. A 10-incher clearly shows the arms, a bright disk of light surrounding the nucleus, and two knots of light in the arms (giant emission nebulae). NGC 2403 is an object that deserves scrutiny on the best nights — you may be surprised at the detail you can see.

Photo by Martin C. Germano

30. M46 (NGC 2437)
Open cluster
R.A. and Dec.: 7h41.8m -14°49'
Con.: Puppis
Mag.: 6.1
Size: 20'

NGC 2438
Planetary nebula
R.A. and Dec.: 7h41.8m -14°44'
Con.: Puppis
Mag.: 11.0
Size: 73" by 68"

Rarely does a deep-sky observer get such good luck that two unrelated objects lie within the same telescopic field of view. But this is exactly the case with the bright star cluster M46 and the faint planetary nebula NGC 2438. Long ago astronomers suspected that the planetary nebula might have formed from one of the stars in the cluster, but this is clearly not the case; modern studies show the planetary is a foreground object lying at a distance of about 3,000 light-years, just 60 percent of the way to the cluster.

The M46/NGC 2438 field is easy to find. Start at Sirius (Alpha Canis Majoris) — the sky's brightest star — and move 4° east-northeast to Gamma Canis Majoris. Next, aim your scope at a spot 10° east, and you'll see a field with two bright stars separated by 2.5° that are oriented northeast-southwest. The M46/NGC 2438 pair lies just above an imaginary line drawn through this pair.

M46 is a rich cluster composed of numerous faint stars. More than 200 stars are visible in the cluster with a small backyard telescope, and many more field stars appear in a wide field. The planetary lies on the northeastern edge of the brightest part of the star cluster and is easily visible as a ring-shaped glow in small scopes. The nebula's central star, a 16th-magnitude object, is challenging in most backyard scopes.

Photo by Bill Iburg

31. NGC 2516
Open cluster
R.A. and Dec.: 7h58.3m -60°52'
Con.: Carina
Mag.: 3.8
Size: 30'

Southern hemisphere observers have a treat in the open star cluster NGC 2516 in Carina. Perched on the edge of the bright part of the southern Milky Way, this bright star group is easily visible with the naked eye and extends over an angular diameter equal to that of the Full Moon. More than 100 stars belong to this rich cluster, including three prominent 8th-magnitude double stars (h4027, h4031, and I 1104).

NGC 2516 lies in southwestern Carina, just 3° southwest of the blazing star Epsilon Carinae. This cluster is remarkably pretty in binoculars because of the large number of bright white stars involved, but it is also a fascinating object to explore with telescopes at low or high magnification.

Photo by Jim Barclay

32. Zeta Cancri
Triple star
R.A. and Dec.: 8h12.2m +17°39'
Con.: Cancer
Mags.: 5.6, 6.0, 6.2
Seps.: 1.0", 6.1"

Many binary stars lie scattered across the springtime sky, but few triple stars are easily observed with small telescopes. One of these is Zeta Cancri, a 5th-magnitude star whose magnitude 5.6, 6.0, and 6.2 components can be split easily in any small telescope. The star was thought to be double after its discovery in 1756, until William Herschel found the third component. All three stars are yellow and the separations are currently 1.0" and 6.1", making all stars in the system visible in virtually any backyard telescope.

As you peer into the eyepiece, consider some amazing facts. The Zeta Cancri system lies about 70 light-years away, so the photons you are now capturing with your eyes left the stars about the time of the Roaring Twenties. The actual gap in space between the two brightest stars is 1.8 billion miles, approximately the distance between the Sun and Uranus. The faintest star shows irregular orbital motion over time, leading astronomers to believe it has a massive, invisible companion.

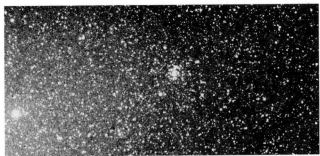
Photo by Robert Provin and Brad Wallis

33. The Beehive Cluster (M44, NGC 2632)
Open cluster
R.A. and Dec.: 8h40.1m +19°59'
Con.: Cancer
Mag.: 3.5
Size: 95'

One of the most easily visible deep-sky objects in the spring sky is the star cluster M44 in Cancer. A sprawling, luminous patch to naked-eye observers, this object was undoubtedly observed in antiquity. As long ago as 130 B.C. Hipparchus dubbed it the "little cloud," and even earlier Aratus called it the "little mist," using its visibility as an indicator of clear weather. The more recent popular names attached to M44 are the Beehive Cluster, or Praesepe.

Because of its fuzzy appearance to naked-eye viewers, the nature of M44 remained elusive until the invention of the telescope. Today we can repeat the observations of the early telescopists and see — with small binoculars or the smallest of telescopes — that M44 consists of several hundred stars spread over a 3° diameter. The cluster is easy to locate, lying in the center of the constellation and surrounded by a trapezoid of bright stars (Delta, Theta, Gamma, and Eta Cancri).

The brightest stars in M44 are just fainter than 6th magnitude and are an even mixture of white, yellow, and pale orange suns. The four brightest stars in the group form a slightly crushed box in the cluster's center. A bright triangle of stars lies immediately west, and curving lines of bright stars extend outward in all directions. The entire region is covered by a dusting of fainter stars, making the Beehive Cluster an impressive location for telescopic sweeping.

Photo by Jay Anderson

34. NGC 2841
Spiral galaxy
R.A. and Dec.: 9h22.0m +51°58'
Con.: Ursa Major
Mag.: 9.3
Size: 8.1' by 3.8'

Ursa Major is a constellation chock full of galaxies. NGC 2841 is one of Ursa Major's best galaxies in that it is large and bright, and it is a pinwheel-like spiral whose form is visible with large amateur telescopes. NGC 2841 can be found southwest of the bowl of the Big Dipper just within the same low-power field as the 6th-magnitude double star 37 Lyncis. At magnitude 9.3, NGC 2841 is visible as an elongated smudge of gray light in a 2-inch scope. A 6-inch instrument shows this object reasonably well, revealing a tiny, bright nucleus enveloped in an oval haze measuring 5' by 2'. With a large backyard scope — in the 16-inch range — you'll be able to detect a mottled, "grainy" appearance in the disk-like spiral arms. This is due to large amounts of dark matter within the arms that are visible with averted vision.

Photo by Martin C. Germano

35. NGC 2903
Spiral galaxy
R.A. and Dec.: 9h32.1m +21°30'
Con.: Leo
Mag.: 8.9
Size: 13' by 6.6'

Leo's NGC 2903 is roughly comparable in brightness to NGC 2841 but is more impressive because of its large angular size. Nestled within the northwestern part of Leo, immediately west of the constellation's Sickle asterism, NGC 2903 is visible in binoculars. Sweep the binoculars from the Sickle over to the 4th-magnitude star Lambda Leonis — the galaxy lies 2° south of this star.

In a telescope, NGC 2903 appears as a large oval haze with a slightly brighter middle. A 10-inch scope shows a series of bright patches within the inner arms that almost passes for a bar shape. Telescopes 16-inches and larger in aperture reveal the faint pattern of this galaxy's many spiral arms, intertwined with thin veins of dark matter.

Photo by Paul Roques

36. M81 (NGC 3031)
Spiral galaxy
R.A. and Dec.: 9h55.6m +69°04'
Con.: Ursa Major
Mag.: 6.8
Size: 26' by 14'

M81 is one of two related galaxies located in the northwesternmost reaches of Ursa Major. To locate this Sb-type spiral, start with Dubhe (Alpha Ursae Majoris) and move 10° west to the 4th-magnitude double star 23 Ursae Majoris. From there, swing your telescope 7° north to the 5th-magnitude star 24 Ursae Majoris. The field containing M81 (and several other galaxies) is just 2.5° southeast of this star.

The greatest attraction of M81 is that it comes with a bright companion galaxy in the same field, M82 (see below). M81 is a 7th-magnitude spiral with an unusually high surface brightness nucleus and hub (both of which are visible in small scopes) and low surface brightness arms. When you observe M81 be sure to notice the two double stars, Σ1386 and Σ1387, that lie in the same field.

M81 and M82 were discovered by Johann E. Bode at the Berlin Observatory in 1774. Observers have been fascinated with M81, describing it as "a fine bright oval nebula of white color" (William H. Smyth); "a spiral which is not well-defined at its boundary and is surrounded by rings of nebulous matter" (Isaac Roberts); and a "nebula [that] is a little oval, the center clear, and can be seen well in an ordinary telescope" (Charles Messier).

Photo by Kim Zussman

Photo by Kim Zussman

37. M82 (NGC 3034)
Irregular galaxy
R.A. and Dec.: 9h55.8m +69°41'
Con.: Ursa Major
Mag.: 8.4
Size: 11' by 4.6'

A mere 38' north of M81 lies its bright companion, M82. This galaxy is a different animal, however: it is an edge-on peculiar system that has shown evidence of violent upheavals within its nucleus. Astronomers have recently produced images showing this galaxy ejecting material at extraordinarily high speeds; the conventional wisdom suggests a large black hole may inhabit M82's center.

M82 is an exciting galaxy to observe. As is the case with most edge-on galaxies, M82 has a high surface brightness that permits using high powers while retaining good contrast and detail in the image. In small scopes M82 appears like a silvery sliver of light. An 8-inch scope on a good night reveals subtle dark blotches on the central part of the galaxy — these are giant clouds of dust along the edge of M82's disk.

38. The Ghost of Jupiter (NGC 3242)
Planetary nebula
R.A. and Dec.: 10h24.8m -18°38'
Con.: Hydra
Mag.: 7.8
Size: 45" by 36"

Planetary nebulae were named so because to early observers they resembled the disks of planets. A planetary nebula in Hydra, NGC 3242, looks so much like a large planet it has acquired the nickname the Ghost of Jupiter. This ghostly greenish object can be found in a blank region of sky 2° south of Mu Hydrae, in a lonely field accompanied by only a solitary 7th-magnitude star.

Finding this obscure field pays off. The Ghost of Jupiter immediately stands out as nonstellar in small scopes and, once located and viewed at high magnification, shows detail within its disk. The Ghost has a "cat's eye" appearance that reminds one of the CBS television logo. An 11th-magnitude central star is surrounded by a bright, oval-shaped disk that is in turn enveloped in a fainter, circular halo of light. The entire structure of this object, different from most planetaries, is visible in an 8-inch telescope on a dark night.

Photo by Jack Marling

Photo by Gregg D. Thompson

39. IC 2602
Open cluster
R.A. and Dec.: 10h43.2m -64°24'
Con.: Carina
Mag.: 1.9
Size: 50'

Sometimes called the "Southern Pleiades" because of its astonishing size and magnitude, IC 2602 is one of the most brilliant star clusters in the sky. Unfortunately for most northern hemisphere viewers, it lies rather deep in the southern part of Carina and is invisible. For those who can see it, this memorable naked-eye object shines with the light of a 2nd-magnitude star and covers nearly two Moon diameters. Much of the cluster's brightness comes from the 3rd-magnitude star Theta Carinae, located in the heart of the cluster. Theta is surrounded by a cloud of stars between magnitudes 5 and 12, nearly 100 in all.

To locate the Southern Pleiades, start at Acrux (the brightest star in the Southern Cross) and move 15° west. A fascinating region for binoculars or wide-field telescopes, the field of IC 2602 should not be missed when observing in the southern hemisphere.

40. The Eta Carinae Nebula (NGC 3372)
Emission nebula
R.A. and Dec.: 10h43.8m -59°52'
Con.: Carina
Mag.: —
Size: 120' by 120'

A simple move 5° north of IC 2602 will bring into view the biggest and brightest emission nebula in the our Galaxy. Though northern observers tend to think of the Orion Nebula as unsurpassed in splendor, the Eta Carinae Nebula, a resident of the deep southern hemisphere sky, makes M42 look second-class. Magnitudes for emission nebulae are not well known — the surface brightnesses of the objects vary enormously over their extensive areas — but certainly Eta Carinae is substantially brighter than its cousin in Orion. It is also much larger, spanning 120' by 120' relative to M42's 90' by 60'. What does all of this mean? If you're able to view the Eta Carinae Nebula you'll get the best view of a stellar birthplace you possibly can.

The curious object that lies inside this nebula is the star Eta Carinae, an odd nebular variable of enormous stellar mass. First recorded by Edmond Halley in 1677, the star irregularly fluctuated in brightness until reaching 2nd magnitude (a two-magnitude jump) in 1730, falling two magnitudes by 1782, and surging in brightness at the turn of the nineteenth century. This peculiar variability continued until the star reached a maximum of magnitude -0.8 in 1843. During the 1860s the star fell below naked-eye visibility, where it has stayed since.

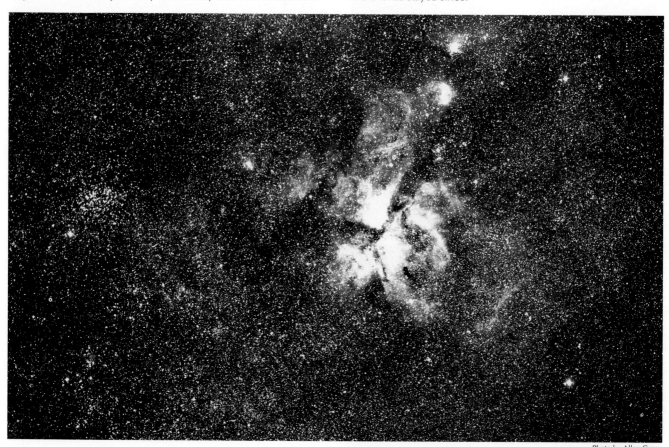

Photo by Allan Green

41. The Owl Nebula (M97, NGC 3587)
Planetary nebula
R.A. and Dec.: 11h14.8m +55°02'
Con.: Ursa Major
Mag.: 9.9
Size: 3.4' by 3.3'

The Owl Nebula in Ursa Major is a planetary nebula found in the bowl of the Big Dipper. Elementary to find, the Owl lies 2° southeast of the star Merek (Beta Ursae Majoris), the bottom front star in the dipper's bowl. (The edge-on galaxy M108 lies 1° northwest of the Owl, just outside a low-power field of view.)

Although this nebula has a rather low surface brightness, it is observable in small telescopes if the night is dark and transparent. After finding the nebula with low power, switch to moderately high magnification (about 150x) and carefully scan over the surface of the Owl. If you're using an 8-inch or larger scope and the seeing is steady, you should faintly see the nebula's central star (which glows softly at 14th magnitude) and two subtle "holes" in the disk of the nebula. These holes gave rise to the name Owl Nebula. Some observers have reported seeing a slight blue-green color in this nebula, but its surface brightness is low enough that most viewers see the nebula as pale white.

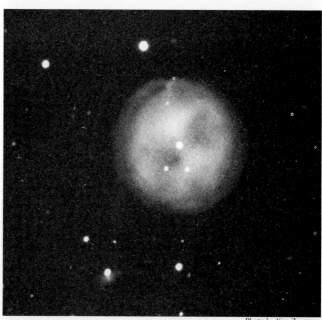

Photo by Kim Zussman

42. M65 (NGC 3623)
Spiral galaxy
R.A. and Dec.: 11h18.9m +13°06'
Con.: Leo
Mag.: 9.3
Size: 10' by 3.3'

M66 (NGC 3627)
Spiral galaxy
R.A. and Dec.: 11h20.2m +13°00'
Con.: Leo
Mag.: 9.0
Size: 8.7' by 4.4'

As fine a pair of galaxies as you could hope for awaits your gaze in the constellation Leo. Direct your telescope to the correct position to see them by aiming toward 3rd-magnitude Theta Leonis, the star marking part of the triangle that constitutes the hindquarters of Leo. Move south 3° and you'll see a prominent north-south row of three 5th-7th magnitude stars. From the brightest (northernmost) of these stars, swing your scope 1° east and you'll immediately see the galaxy field. (The bright edge-on galaxy NGC 3628 lies just north of M65 and M66).

M65 and M66 lie together in the same field, separated by a mere 21'. Both galaxies are Sb-type spirals, although M66 is inclined such that its arms are more easily visible in amateur telescopes (it's brighter, too). More elongated and with a far more prominent disk shape, M65 consists of a bright center surrounded by a halo of light. On nights of good seeing observers with telescopes of 10-inches or more aperture may see dark material in the arms of M66.

Photo by K. Alexander Brownlee

43. M99 (NGC 4254)
Spiral galaxy
R.A. and Dec.: 12h18.8m +14°25'
Con.: Coma Berenices
Mag.: 9.8
Size: 5.4' by 4.8'

One of many galaxies in Coma Berenices, M99 is notable because of its easily visible spiral arms. A typical Sc-type spiral, M99 is essentially face-on to our line of sight and has relatively high surface brightness spiral arms. Because of this, the bright, oval galaxy reveals significant amounts of detail on moonless nights when the sky is particularly transparent.

M99 was discovered in 1781 by the French comet hunter Pierre Mechain, who found it 1° southeast of the 5th-magnitude star 6 Comae Berenices. (The galaxy itself lies in a field with a lone 7th-magnitude sun.) Small scopes and binoculars show M99 as a circular glow surrounding a bright center. A good 6-inch scope is required to see the galaxy's spiral arms and a bright, ball-like nucleus. After you observe M99, scan the surrounding field, which is full of dimmer galaxies. M98, another spiral, lies 1.3° to the northwest, on the other side of 6 Comae Berenices.

Photo by Lee C. Coombs

44. M106 (NGC 4258)
Spiral galaxy
R.A. and Dec.: 12h19.0m +47°18'
Con.: Canes Venatici
Mag.: 8.3
Size: 18' by 7.9'

M106 is an Sb-type spiral galaxy located in Canes Venatici not far from the bowl of the Big Dipper. To find it, start at Phecda (Gamma Ursae Majoris), the southeasternmost star in the dipper's bowl. Move south 6° to the 4th-magnitude star Chi Ursae Majoris, then east slightly more than 5° to a lone 6th-magnitude star. M106 lies less than 1° west of this solitary sun.

Although the published dimensions of this galaxy show it is an enormous 18' long, the bright disk portion is about half that length. (The spiral arms that extend to 18' in length have a low surface brightness and are difficult to see in backyard scopes.) A 4-inch telescope shows this object as a bright oval nebulosity 6' long with a brighter center. On an exceptionally dark night an 8-inch scope reveals some detail in M106, showing a bright, condensed nucleus surrounded by a halo of nebulosity about 10' long. Large amateur scopes reveal some mottling in the disk component of M106, caused by clouds of dark material between the arms, and with averted vision a hint of the outer spiral arms.

Photo by Kim Zussman

45. M100 (NGC 4321)
Spiral galaxy
R.A. and Dec.: 12h22.9m +16°49'
Con.: Coma Berenices
Mag.: 9.4
Size: 6.9' by 6.2'

The largest and brightest spiral in the Virgo Cluster, M100 is chiefly remembered for producing a bright supernova in 1979. It is an unusually interesting galaxy to observe, however, because of its relatively high surface brightness, even without a supernova.

M100 might be more frequently observed if it didn't lie in the thick of the Virgo Cluster surrounded by hordes of other galaxies. Despite the crowded environs, it is relatively easy to locate. Start your star-hop at the 2nd-magnitude star Denebola, the easternmost bright star in Leo. Move your telescope or binoculars 7° east and slightly north to the 5th-magnitude star 6 Comae Berenices (the jumping off point for M99, discussed earlier). The star 6 Comae Berenices forms a line with two slightly fainter stars that extend over 2° to the northeast. M100 lies 1° past the last star in this row.

M100 is a viewing treat. Even a 3-inch telescope shows the galaxy's bright nucleus, which measures about 2' across, enveloped in a faint haze that glows with a subtle greenish hue. Eight-inch scopes reveal a hint of the galaxy's spiral structure, a feature visible with direct vision in 12-inch and larger instruments.

Photo by Martin C. Germano

Photo by Alan Dyer

46. The Coma Star Cluster (Melotte 111)
Open cluster
R.A. and Dec.: 12h25.1m +26°06'
Con.: Coma Berenices
Mag.: 1.8
Size: 275'

Obvious to the naked eye on any springtime night, the great collection of stars that makes up most of the constellation Coma Berenices is not a random pattern but a physically bound star cluster. Altogether spanning more than 4° in diameter, this bright star group is a joy to scan with binoculars or a small telescope equipped with wide-field eyepieces. Many of the bright stars that make up this cluster are doubles, notable examples being 12 Comae Berenices, 17 Comae Berenices, and 30 Comae Berenices. The region is littered with a backdrop of galaxies, including the spectacular edge-on spiral NGC 4565 (see object #47).

The brightest stars in the group are 12 Comae Berenices (magnitude 4.8, yellow), 14 Comae Berenices (5.0, white), 16 Comae Berenices (5.0, white), 13 Comae Berenices (5.2, white), 21 Comae Berenices (5.5, white), and 22 Comae Berenices (6.3, white). More than 30 stars are physically associated with this group, which lies 250 light-years away and, like all open clusters, will slowly shred apart as it orbits the center of the Galaxy.

Photo by Kim Zussman

47. NGC 4565
Spiral galaxy
R.A. and Dec.: 12h36.4m +26°00'
Con.: Coma Berenices
Mag.: 9.5
Size: 16' by 2.8'

"It is a very beautiful object, very well seen in the finding eyepiece; the whole nebula is much broader at nucleus than elsewhere, narrowing off suddenly, and the nucleus projects forward into the dark space." So William Parsons, the Third Earl of Rosse, described the galaxy NGC 4565 in 1855. Of course then Rosse had no idea that this "spiral nebula" is a distant galaxy, but his large telescopes showed him the object as nicely as those owned by amateurs today.

NGC 4565 is the classic edge-on spiral galaxy. At magnitude 9.5 it is bright enough to be visible in a 2-inch telescope, and the galaxy's high surface brightness permits an impressive view in a 4-inch instrument. Consisting of a broad, bright, ball-like nucleus and two tapering sides, NGC 4565 appears like a sliver of glass made visible by ghostly illumination. Telescopes 6 inches and larger easily show the galaxy's broad dark lane, composed of dust lying along the equatorial plane and visible against the brightly glowing light from millions of stars. NGC 4565 is a normal Sb-type galaxy, much like our Milky Way. As such, some viewer in a distant galaxy may see our home Galaxy much as we see NGC 4565.

48. The Sombrero Galaxy
(M104, NGC 4594)
Spiral galaxy
R.A. and Dec.: 12h40.0m -11°37'
Con.: Virgo
Mag.: 8.3
Size: 8.9' by 4.1'

Another splendid edge-on galaxy lies in the constellation Virgo, just over Virgo's border with the more southern group of stars called Corvus. This one is different from NGC 4565, however. M104, popularly known as the Sombrero Galaxy, is a big fat Sa-type galaxy that shows characteristics of ellipticals as well as spirals. The galaxy has a bright disk component but also a massive spherical halo of gas and dust that saturates the space surrounding the disk. The galaxy is not oriented as precisely edge-on as NGC 4565, which conspires to give M104 the general appearance of a spaceship from a B-movie sci-fi thriller.

But unlike spaceships, the Sombrero Galaxy is real. To locate it, start by aiming your scope toward the bright stars Delta and Eta Corvi, the objects that mark the northeastern corner of the Corvus trapezoid. Move 5° northeast and you'll come to a solitary 5th-magnitude double star. From there, move 1.5° north and slightly west and you'll be centered on M104's position.

Small telescopes show the Sombrero as a bright bulge of light that is essentially oval in shape. A good 6-inch scope reveals a bright center, a sharply defined disk, traces of the faint halo that surrounds the disk, and a broad, obvious dust lane running along the galaxy's equator.

Photo by Kim Zussman

49. Porrima (Gamma Virginis)
Double star
R.A. and Dec.: 12h41.7m -1°27'
Con: Virgo
Mags.: 3.5, 3.5
Sep.: 2.9"

One of the easiest equal-magnitude double stars for small telescopes is Porrima, a star surrounded by dozens of bright galaxies in the springtime sky. While most deep-sky observers are typically observing galaxies in the spring (rather than double stars), it is well worth taking a break from the faint fuzzies to see Porrima.

This star is easy to find, lying 14.5° northwest of Spica (Alpha Virginis) and an equal distance south of the heart of the Virgo Cluster of galaxies. While most binary stars have components different in color and brightness, Porrima's stars are essentially identical. Both are white stars of spectral type F0 and both shine at magnitude 3.5. Although the pair's separation varies considerably over time, the stars are currently divided by about 2.9". (The separation was last at maximum in 1920, when the gap between the stars was 6.2".)

Porrima is a close star, lying only 32 light-years away. The twin stars are considerably brighter than the Sun, each having a luminosity seven times greater. Porrima is one of the most observed of the springtime doubles, having become a popular object not long after its discovery in 1718.

50. M94 (NGC 4736)
Spiral galaxy
R.A. and Dec.: 12h50.9m +41°08'
Con.: Canes Venatici
Mag.: 8.1
Size: 11' by 9.1'

M94 is a high-surface-brightness, tightly wound spiral galaxy that is visible in the smallest backyard telescopes. It is remarkably easy to find, lying south of the handle of the Big Dipper in Canes Venatici. From the 4th-magnitude star Beta Canum Venaticorum, simply move your scope 3° east and you'll be aimed precisely at M94. (The galaxy forms an isosceles triangle with Alpha and Beta Canum Venaticorum, by the way.)

When you observe M94, crank up the magnification to about 200x and take a look at the galaxy's center. You'll note something for which M94 is well known, a brilliantly intense center. Surrounding this central glow is a larger circular halo of light from the galaxy's spiral arms. The detail in M94's arms is so fine that little of it can be seen in amateur scopes, however. Even the largest backyard scopes show the bright central region enveloped in a soft, nebular light about 8' in diameter.

Photo by John Sanford

Photo by Martin C. Germano

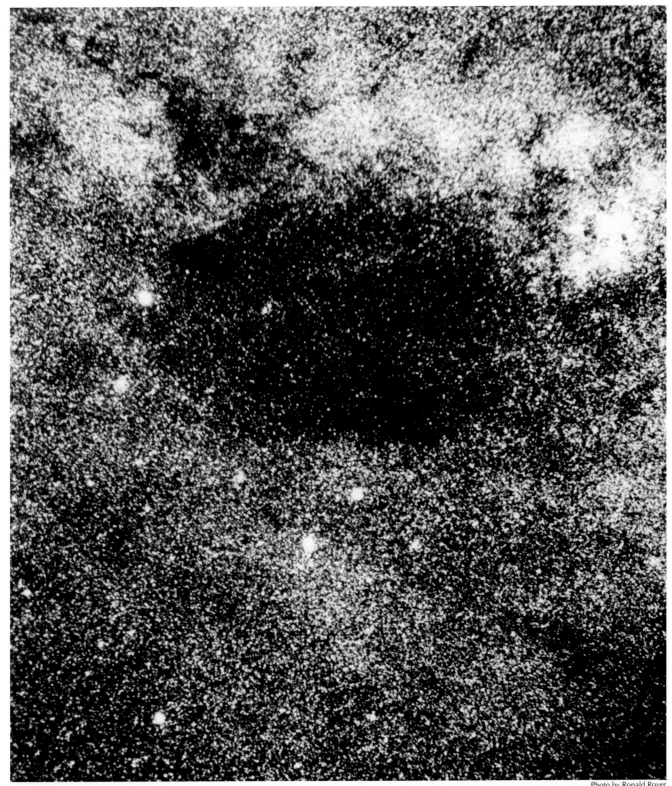

Photo by Ronald Royer

51. The Coal Sack
Dark nebula
R.A. and Dec.: 12h53.0m -63°00'
Con.: Crux
Mag.: —
Size: 400' by 300'

A striking feature of the southern sky for naked-eye observers is an immense dark nebula called the Coal Sack. It appears as a giant "hollow" devoid of any Milky Way glow immediately southeast of the Southern Cross, the bright asterism in the constellation Crux. (The Coal Sack extends into part of Musca and Centaurus, too.) Measuring 7° by 5° in extent and lying right smack in the galactic plane, the Coal Sack is unavoidable as an eye-grabbing feature of the southern sky. It is probably the nearest dark nebula to us, lying perhaps 500 light-years away and spanning a region 70 light-years across.

Scanning the Coal Sack with a pair of binoculars will provide a wonderful demonstration about dark clouds in the Galaxy. Originally thought to be a "hole" in the Milky Way, the Coal Sack — and other dark nebulae — are now known to be opaque clouds of fine dust particles that block starlight from behind. As you look at the richness of Crux itself and the paucity of bright stars within the Coal Sack — only 11 stars brighter than 9th magnitude inhabit it — you'll see how obscuring these dust clouds can be.

52. The Jewel Box Cluster (NGC 4755)
Open cluster
R.A. and Dec.: 12h53.6m -60°20'
Con.: Crux
Mag.: 4.2
Size: 10'

Just north of the Coal Sack — between it and the bright star Mimosa (Beta Crucis) — is a rich open cluster designated NGC 4755. Visible to the naked eye, this cluster so dazzled the 19th-century observer John Herschel that he claimed it appeared like jewelry. Thus the cluster's popular name, the Jewel Box Cluster, was born.

As open clusters go, the Jewel Box sparkles with many bright stars in a small diameter of only 10'. Altogether, 50 bright stars and numerous fainter ones lie within this area. The brightest stars are all blue-white supergiants shining between magnitudes 5.8 and 11.4 — with the exception of one red supergiant, a magnitude 7.6 object. The overall effect is of a field blazing with the light from distant, powerful stars. The brightest stars in the group are arranged in a V-shape, with many fainter ones scattered more or less uniformly over and around this pattern.

The distance to the Jewel Box Cluster is about 7,700 light-years, which makes the central part of the cluster 25 light-years in diameter.

Photo by Chris Floyd

Photo by Kim Zussman

53. The Blackeye Galaxy (M64, NGC 4826)
Spiral galaxy
R.A. and Dec.: 12h56.8m +21°41'
Con.: Coma Berenices
Mag.: 8.5
Size: 9.3' by 5.4'

A few galaxies offer small telescope observers the chance to see dark dust clouds from a vantage point millions of light-years away. M64 in the constellation Coma Berenices is one of these. Aside from being a rather normal, bright Sb galaxy, M64 contains a huge dark patch that floats in front of half the galaxy's disk. Prominently visible in small telescopes, this dark patch has given rise to the nickname Blackeye Galaxy.

To find M64, start by locating the 4th-magnitude double star Alpha Comae Berenices, the brightest star in the constellation. Move 5° northwest and you'll come to a 5th-magnitude double, 35 Comae Berenices. M64 lies just over 1° northeast of this star.

In small scopes, M64 appears as a bright oval nebulosity with a sharply bright center and a thin patch of darkness on one side of the nucleus. An 8-inch scope reveals a faint, outer halo of light from the galaxy's spiral arms which greatly increases the galaxy's size, and shows an arc shape in the cloud of dust.

Photo by Paul Roques

54. M63 (NGC 5055)
Spiral galaxy
R.A. and Dec.: 13h15.8m +42°02'
Con.: Canes Venatici
Mag.: 8.6
Size: 12' by 7.6'

M63 is a bright Sc-type spiral that is famous for its wide, knotty spiral arms. Visible in binoculars as a faint smudge of light, M63 is a hallmark of the spring sky that should not be missed by telescope users. The galaxy's bright nucleus and clumpy, irregular outer arms form a memorable sight in large backyard instruments.

To locate M63, start with the 3rd-magnitude star Alpha[1,2] Canum Venaticorum. From there, move 4° northeast to a shield asterism consisting of five bright stars, including 18, 19, 20, and 23 Canum Venaticorum. The galaxy can be found about 1.5° north of 6th-magnitude 19 Canum Venaticorum.

With its great brightness and large dimensions, M63 is visible in any telescope. But because its outer spiral arms have a low surface brightness, the galaxy appears best in scopes 8 inches and larger in aperture. Such a telescope shows a condensed, ball-like glow in the galaxy's center, a bright disk of light surrounding it, and a much fainter oval halo of nebulosity. On nights of steady seeing you'll perceive a mottled effect in the galaxy's arms, which results from numerous dust clouds scattered throughout the galaxy.

55. Mizar (Zeta Ursae Majoris)
Double star
R.A. and Dec.: 13h23.9m +54°56'
Con.: Ursa Major
Mags.: 2.3, 4.0
Sep.: 14.4'

One of the standard tests for reasonably adequate visual acuity is to look at the handle of the Big Dipper, focus in on the middle star in the handle, and see two distinct stars separated by a gap of inky black sky. This is not difficult, and anyone in a dark sky with decent vision should have no problem doing it. The two stars involved are Mizar (the bright one) and Alcor, and they are separated by 11.8'. However, Alcor and Mizar merely masquerade as a pair: they are not physically related, but simply lie along the same line of sight in our sky. Although they are in the same vicinity of space, the two stars lie 1.5 trillion miles apart.

There is a true binary star in the area. It is Mizar itself, which when viewed in a telescope is an obvious and pretty pair of stars that was in fact the first double star discovered, back in 1650, by Giovanni Riccioli. The two stars are separated by just over 14' and shine brilliantly at magnitudes 2.3 and 4.0, so that Mizar is a cleanly resolved and easily enjoyed sight in any instrument. As you gaze at these two sparkling stars, consider that the stars are about 35 times more luminous than the Sun and the gap of space you see between them represents 35 billion miles (or nearly five times the distance between the Sun and Pluto).

Photo by Robert Provin and Brad Wallis

56. Centaurus A (NGC 5128)
Lenticular galaxy
R.A. and Dec.: 13h25.5m -42°01'
Con.: Centaurus
Mag.: 7.0
Size: 18' by 14'

Centaurus A is one of the strangest galaxies known. Long thought to be the result of the collision of two galaxies, this odd creature is essentially a giant elliptical galaxy that shows signs of containing a disk structure as well. It is highly energetic, emitting focused beams of radio energy and probably harboring a black hole deep in its "central engine." Centaurus A also presents one of the thickest and most easily visible equatorial dust lanes of any galaxy. It is high on the checklists of all galaxy observers, whether or not they can travel to the southern latitudes to see it.

This galaxy is one of the brightest in the sky and so is easy to find. Start at the prominent triangle of 3rd-magnitude stars Psi, Nu, and Mu Centauri. Move 4° west and slightly south from Mu Centauri, and the galaxy will arrive in your low-power eyepiece.

Small scopes show a hazy ball of nebulosity lying in a rich star field. With reasonably high magnification on a 2-inch scope, the equatorial dust lane is obvious. An 8-inch scope shows a bright disk of luminous material 8' across bisected by the wide dark lane and surrounded by a faint, low surface brightness glow. Telescopes 16 inches or larger show that the dust lane broadens at the galaxy's edges and that it contains small clumps of glowing material, especially near the galaxy's center.

Photo by Jim Barclay

57. Omega Centauri (NGC 5139)
Globular cluster
R.A. and Dec.: 13h26.8m -47°29'
Con.: Centaurus
Mag.: 3.5
Size: 36'

Not far from Centaurus A lies a globular cluster well worth observing by those who are far enough south to see it. The largest and brightest globular cluster in the sky, Omega Centauri is a treasure that is impressive to first-time viewers even in binoculars.

Visible to the naked eye as a hazy "star," Omega Centauri is easily located. Start at the 3rd-magnitude star Zeta Centauri and move 4.5° west to the cluster.

When you get there, you'll see that Omega Centauri lacks a brilliant, ball-like nucleus but instead offers a huge cloud of faint stars covering an area larger than the Full Moon. A 3-inch scope shows a luminous glow that is fairly evenly illuminated, although the center is somewhat brighter than the outer parts. A 6-inch scope resolves the cluster's brightest stars and shows the center as a mottled, hazy light. Omega Centauri is an unforgettable sight in any telescope, although scopes 8 inches or larger in aperture show many of the individual stars resolved, yielding a three-dimensional effect. Large backyard scopes show a sparkling field of pinpoint stars floating in front of a giant unresolved mass of greenish light. Regardless of the telescope used, Omega Centauri is a sight not easily forgotten.

Photo by Jean Dragesco

A Portfolio of the Finest Deep-Sky Objects

The Lagoon and Trifid nebulae. photo by Tony Hallas and Daphne Mount

The Orion Nebula, photo by Randall R. Pfeiffer

Galaxy M101, photo by
Tony Hallas and Daphne Mount

The Whirlpool Galaxy, photo by
Tony Hallas and Daphne Mount

The Double Cluster, photo by Mace Hooley

The North America Nebula, photo by Vance C. Tyree

Centaurus A, photo by Jack B. Marling

The Eagle and Omega nebulae, photo by Matthew Wilson

The Winter Milky Way, photo by Mark J. Coco

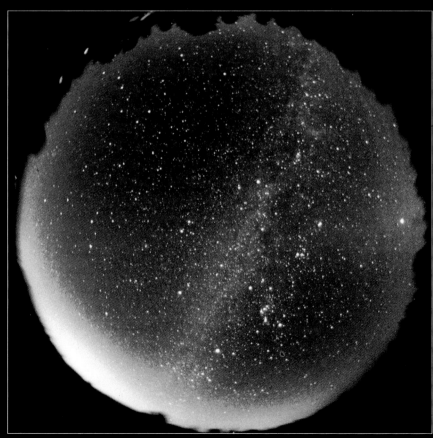

The Andromeda Galaxy, photo by Tony Hallas and Daphne Mount

The Pleiades, photo by Philip J. Lambert

The Horsehead Nebula, photo by Tony Hallas and Daphne Mount

Galaxy M33, photo by Larry Myers

The Cocoon Nebula, photo by Bill Iburg

The Large Magellanic Cloud, photo by Simon Tsang

The Rosette Nebula, photo by Mike Simmons

The Eta Carinae Nebula, photo by Jim Barclay

Omega Centauri, photo by Jack Newton

Galaxy NGC 253, photo by Jack B. Marling

The S Nebula, photo by Michael Stecker

The Veil Nebula, photo by Jack Newton

Photo by Kim Zussman

58. The Whirlpool Galaxy
(M51, NGC 5194)
Spiral galaxy
R.A. and Dec.: 13h29.9m +47°12'
Con.: Canes Venatici
Mag.: 8.4
Size: 11' by 7.8'

Because of its unique appearance, the Whirlpool Galaxy in Canes Venatici has become one of the universal icons of astronomy. The Whirlpool actually consists of two galaxies, a large, Sb-type spiral (the Whirlpool Galaxy itself, M51) and a small, high-speed interloper, NGC 5195, that has passed by M51 and is stripping material from the larger galaxy. Because of the high surface brightnesses of both galaxies, the Whirlpool provides a chance for small telescope owners to see galaxies interacting.

Star-hopping to the Whirlpool Galaxy begins below the Big Dipper. From Eta Ursae Majoris, the end star of the dipper's handle, move 2.5° west to the 6th-magnitude star 24 Ursae Majoris. From there move south and slightly west an equal distance, and you'll spot the fuzzy light from the Whirlpool Galaxy.

A 4-inch telescope shows the Whirlpool as a fuzzy glow 6' across with a much smaller nebulosity — NGC 5195 — nearly in contact to the northeast. A good 8-inch telescope reveals much more detail, showing the delicate face-on spiral pattern of the main galaxy, the non-circular shape of NGC 5195, several faint foreground stars superposed on M51, and the faint bridge of light connecting M51 with NGC 5195. At high-power slight mottling is visible amid the galaxy's arms, caused by dark matter within the galaxy.

Photo by Martin C. Germano

59. M83 (NGC 5236)
Spiral galaxy
R.A. and Dec.: 13h37.0m -29°52'
Con.: Hydra
Mag.: 8.2b
Size: 11' by 10'

Although it lies close to the horizon for many northern observers, the Sc-type galaxy M83 is an excellent object because of its high surface brightness, knotty spiral arms, and large diameter. The galaxy lies in a rich star field in Hydra, approximately 14° north of the lenticular galaxy Centaurus A (see object #56).

To locate M83, aim your finder scope or binoculars at the 3rd-magnitude star Pi Hydrae. Next, move your scope 5° west to a lone 5th-magnitude double star. From there drop down 3.5° in declination, and you'll be centered on the galaxy.

M83 is a classic Sc-type spiral. It has a small, bright nucleus surrounded by an almost bar-like extension from which the spiral arms protrude. The arms are loose, thin, and knotted with clumps of stars and dark matter. The entire spiral pattern is enveloped in a low surface brightness halo of soft gray light. The galaxy's spiral pattern is visible in 6-inch and larger scopes, and details such as the knotty structure of the arms and ball-like nucleus can be glimpsed in 10-inch and larger instruments.

60. M3 (NGC 5272)
Globular cluster
R.A. and Dec.: 13h42.2m +28°23'
Con.: Canes Venatici
Mag.: 5.9
Size: 16'

M3 is a globular cluster lying in an area of sky saturated with galaxies. Just visible to the naked eye on dark nights, M3 appears as a hazy, bloated "star" in binoculars and reveals its stellar nature to small telescopes operating at moderately high powers. Overall, M3 appears as a sort of miniature Hercules Cluster (see object #66).

Locate M3 by starting at the Coma Star Cluster (see object #46). From the center of the cluster, swing your telescope 5° east to a bright east-west pair of stars, 30 and 31 Comae Berenices. Continue eastward 3.5°, and you'll see a brighter east-west pair, Beta and 41 Comae Berenices. Keep moving eastward along this line 6°, and you'll come to a tight pair of stars separated by only 30'. The northeasternmost "star" of this pair is the cluster M3.

Like most globulars, M3 contains upward of one million stars. With a 6-inch telescope many of these stars are visible as resolved points of light. If the seeing is steady and the night especially dark, a 10-inch scope cleanly resolves the face of the cluster into myriad stars.

Photo by Martin C. Germano

Photo by Kim Zussman

61. M101 (NGC 5457)
Spiral galaxy
R.A. and Dec.: 14h03.3m +54°22'
Con.: Ursa Major
Mag.: 7.7
Size: 27' by 26'

M101 is a face-on spiral galaxy whose light is so spread out that although the galaxy's total brightness is great, it is difficult to see in small instruments. Scopes of 10 inches or more aperture show a wealth of detail in M101 when viewed against a dark sky, because the galaxy's Sc-type arms are filled with knots, clumps, and pockets of dust.

Star-hopping to M101 begins by finding Mizar, the middle star in the handle of the Big Dipper (see object #55). Extend a line from Mizar through Alcor, its fainter companion, and continue onward for 2° until you reach the 5th-magnitude star 81 Ursae Majoris. Now follow the string of similarly bright stars southeastward to 83, 84, and 86 Ursae Majoris. A 2°-long line of three 7th-magnitude stars aligned north-south can be found just east of 86 Ursae Majoris. Draw an imaginary line from 86 through the center 7th-magnitude star and extend it for 1°. This puts you in the position of M101.

Because of M101's low surface brightness, a 4-inch telescope simply shows a large, fuzzy circular glow. A good 6-incher begins to show some detail on the darkest nights, revealing a small central glow surrounded by a halo of light one-third the size of the Moon. Scopes 10 inches and larger show the galaxy's delicate spiral pattern and reveal irregularities in M101's arms.

62. Xi Bootis
Double star
R.A. and Dec.: 14h51.4m +19°06'
Con.: Bootes
Mags.: 4.7, 7.0
Sep.: 7"

Xi Bootis is a colorful, pretty double star for small telescopes. Consisting of magnitude 4.7 and 7.0 suns, this star has a total light output of magnitude 4.5 and is therefore visible to the naked eye as a single stellar glow. Look for Xi Bootis in the southeastern part of Bootes approximately 8° east of Arcturus (Alpha Bootis).

After aiming a low-power ocular at this system, you'll be enchanted by the colors of the stars. The primary shines with a rich, golden yellow hue, and the secondary — about 7" distant — glows with an eerie, reddish-violet light. The actual distance between the stars is typically about three billion miles, and the system lies a mere 22 light-years away. It was discovered as double in 1780 by the English observer William Herschel.

Photo by Daniel J. McGlaun Jr.

63. M5 (NGC 5904)
Globular cluster
R.A. and Dec.: 15h18.5m +2°05'
Con.: Serpens Mag.: 5.7
Size: 17'

M5 is a globular cluster similar in size and appearance to M3 (see object #60) in Canes Venatici. Located lower in the sky in the sprawling constellation Serpens, M5 is ever-so-slightly larger and brighter, but the difference is barely noticeable at the eyepiece.

Locate M5 by aiming your telescope toward Alpha Serpentis, the 3rd-magnitude beacon of the constellation. Move 7° southeast and you'll see a large triangle of stars consisting of 5, 6, and 10 Serpentis. M5 is the fuzzy "star" that lies in the same low-power field as 5 Serpentis, just northeast of the star.

In the eyepiece of a small telescope, M5 is a large, circular haze peppered with minute yellow and white stars at its edges. Higher powers and larger telescopes show progressively more stars. A 10-inch telescope at 150x cleanly resolves the object across its core, showing many hundreds of individual points of light spread across a smooth background glow.

Photo by Lee C. Coombs

64. R Coronae Borealis
Irregular variable star
R.A. and Dec.: 15h48.6m +28°09'
Con.: Corona Borealis
Mag. range: 5.7-14.8
Period: irregular

Discovered by the English astronomer E. Pigott in 1795, the irregular variable star R Coronae Borealis is a star that normally stays near maximum and occasionally drops nearly seven magnitudes in light output. Because the star can vary without warning and over irregular cycles, it is an object worth monitoring during spring and summer observing sessions.

This star's history reflects its bizarre instability. Pronounced drops in light output have occurred between 1863-1867, 1868-1870, 1872-1874, 1909-1912, 1917-1918, 1938-1940, and 1948-1950. Long periods of inactivity — when R Coronae Borealis simply stayed at 6th magnitude — took place between 1861-1863, 1897-1903, 1924-1934, and 1940-1942.

The star's puzzling behavior most likely results from its carbon-rich atmosphere. Astronomers believe the drops in light output result from the emission of thick clouds of carbon from the star which condense in a cloud of soot and block light from the star itself. As the cloud slowly disperses, the star's normal brightness returns.

Photo by Orien Ernest

Photo by Alfred Lilge

65. M4 (NGC 6121)
Globular cluster
R.A. and Dec.: 16h23.6m -26°32'
Con.: Scorpius
Mag.: 5.8
Size: 26'

Nothing beats an early summer evening of stargazing, as the glowing arch of the Milky Way rises and the stars of Scorpius and Sagittarius dominate the southern sky. One universally favorite object in Scorpius — a must see on summer evenings — is the globular cluster M4, a rich, unusual cluster.

M4 is located near Antares, the brightest star in Scorpius, and is one of the easiest globulars to locate in the sky. From Antares, simply move 1.5° east, and you'll see the hazy glow from M4 in your finder scope.

At a distance of only 7,500 light-years, M4 is one of the closest globular clusters. As such, it is very easy to resolve with backyard telescopes because its brightest stars are much brighter than those in most other globulars. A 4-inch scope at 100x resolves the group's brightest members and shows a feature for which M4 is famous — a bright "bar" of resolved stars crossing the face of the cluster. An 8-inch scope at high power reveals a sparkling field of diamond-like stars that appear to float in front of a gray-green halo of light 20' across.

Photo by Rick Dilsizian

66. The Hercules Cluster (M13, NGC 6205)
Globular cluster
R.A. and Dec.: 16h41.7m +36°27'
Con.: Hercules
Mag.: 5.7
Size: 17'

Not the biggest. Not the brightest. Just the most famous globular cluster, M13 is easily observed by northern hemisphere viewers because it rises high up in the sky on spring and summer evenings. Like M4, it is easy to find. First find the keystone asterism that forms central Hercules. (This figure is composed of Eta, Zeta, Epsilon, and Pi Herculis.) From Eta Herculis, the northwesternmost star in the keystone, move south 2.5°, about a third of the way toward Zeta. This will bring you to the field of M13. (The spiral galaxy NGC 6207 lies 30' northeast of the cluster.)

The Hercules Cluster has an unusually bright, condensed center that is visible in telescopes as small as 2 inches in aperture. The cluster's brightest stars can be resolved with a 4-inch scope. An 8-incher on a dark night provides an impressive view of the cluster: long arcs of stars radiate outward from the center of M13 and spill outside a low-power field. These star chains are especially prominent when the cluster is high overhead and suggest a three-dimensional effect.

67. NGC 6231
Open cluster
R.A. and Dec.: 16h54.0m -41°48'
Con.: Scorpius
Mag.: 2.6
Size: 15'

It's low in the sky but a world-class showpiece. Largely ignored by northern observers because of its southern declination, star cluster NGC 6231 is a sight that shouldn't be missed. Consisting almost entirely of hot O and B stars, this cluster appears as a brilliant gem to naked-eye viewers even though it lies 5,700 light-years away. That is testimony to the enormous intrinsic brightness of the individual stars in this group.

NGC 6231 is located in the so-called Table of Scorpius, just 1° north of the bright stars Zeta[1] and Zeta[2] Scorpii. The best method for observing this group is to use binoculars or a low-power telescope. The twenty brightest stars in this cluster shine with magnitudes of 5.3 to 8.8, yielding a total magnitude of 2.6 squeezed into an area 15' across — half the diameter of the Full Moon.

The most striking thing about NGC 6231 is the high number of bright stars — it seems that several dozen O and B stars shine brightly in a binocular view, although the total number of stars brighter than 9th magnitude is probably 25. Still, this cluster is remarkable in its class and should not be missed.

Photo by Martin C. Germano

68. M92 (NGC 6341)
Globular cluster
R.A. and Dec.: 17h17.1m +43°08'
Con.: Hercules
Mag.: 6.4
Size: 11'

Hercules contains a second globular cluster — not as large or bright as M13 (see object #66) — but well worth the trouble to hunt down and observe. Discovered by Johann E. Bode in 1777, M92 is a rich cluster that is overshadowed by its more famous neighbor. If located elsewhere in the sky, it would be observed more frequently.

To locate M92, start at the other side of the keystone asterism than the spot from which you found M13. This time locate the 2nd-magnitude star Pi Herculis. From there, move 7° north and you'll spot a hazy glow measuring 5' across in the finder scope. This is M92.

In the eyepiece, M92 is a respectable globular cluster showing a fuzzy disk 8' across with a few dozen of the cluster's brightest stars scattered across the surface. The group's central glow is particularly condensed and bright, and this feature is visible in a 3-inch instrument. In large backyard telescopes, the view of M92 is impressive, consisting of hundreds of resolved star images scattered across a bright halo of nebulosity.

Photo by Lee C. Coombs

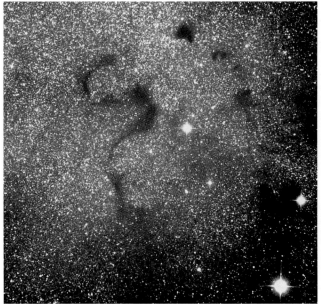
Photo by Martin C. Germano

69. The S Nebula (Barnard 72)
Dark nebula
R.A. and Dec.: 17h23.5m -23°38'
Con.: Ophiuchus
Mag.: —
Size: 4'

Dark nebulae are generally difficult to observe and usually consist of long, sinuous channels that wind through dense star fields. The S Nebula in Ophiuchus, also called the Snake Nebula, is typical only in that it winds through a dense star field. Otherwise, it is atypical of dark nebulae. It is short, dense, and easy to observe.

Start your star-hop to the S Nebula by aiming your finder scope at the bright star Theta Ophiuchi. Be sure you're using a wide-field eyepiece and point the scope 1.5° north-northwest of Theta. On a moonless night, if you're using a 6-inch or larger scope, you should see a decrease in the number of faint stars in parts of the field of view. As your dark adaptation increases and you get used to the star field, you'll see that fewer stars inhabit a region that is S-shaped and extends over about 4'. This is the S Nebula, one of the more opaque dark nebulae in the early summer sky.

70. The Butterfly Cluster (M6, NGC 6405)
Open cluster
R.A. and Dec.: 17h40.1m -32°13'
Con.: Scorpius
Mag.: 4.2
Size: 33'

M6, occasionally known as the Butterfly Cluster, is one of two naked-eye open clusters lying between the bright forms of Scorpius and Sagittarius. Though not as large and bright as the other cluster in the pair, M7 (see object #71), M6 has a distinctive shape that makes it an attractive object for binocular observers.

Easily visible without optical aid, M6 lies 5° north-northeast of Shaula (Lambda Scorpii), the stinger star in the scorpion's tail. (The cluster also lies just 3.5° southwest of the 0° mark along the galactic equator, the point on the sky marking the galactic center.) The cluster's more than 100 stars form a pretty scene. The brightest star, an irregular variable designated BM Scorpii, is an orange star normally shining at magnitude 6.2. Most of the other bright stars are blue or blue-white, and they are positioned such that one could certainly imagine the outspread wings of a butterfly. The cluster covers more than the area of the Full Moon, and together the stars glow at nearly 4th magnitude.

Photo by Lee C. Coombs

71. M7 (NGC 6475)
Open cluster
R.A. and Dec.: 17h53.9m -34°49'
Con.: Scorpius
Mag.: 3.3
Size: 80'

Located just 3.5° southeast of M6 is the much larger and brighter cluster M7, an obvious naked-eye feature of the summer Milky Way. Because it is more than 2.5 Moon diameters across and glows with the light of a 3rd-magnitude star, M7 is a spectacular cluster for binocular viewing or sweeping with a small telescope at low power.

Because of its brilliancy, M7 has been observed throughout history. Ptolemy listed it in his *Almagest*, and many later observers left long series of notes about this object. For all of the observations that preceded his, Charles Messier described it unceremoniously as a "cluster considerably larger than [M6] [that] appears to the naked eye as a nebulosity." Telescopes show several hundred stars scattered across the cluster, about 80 of which shine brighter than 10th magnitude. The group lies about 800 light-years away; the cluster's age is probably 260 million years.

Photo by Martin C. Germano

72. The Trifid Nebula (M20, NGC 6514)
Emission and reflection nebula
R.A. and Dec.: 18h01.9m -23°02'
Con.: Sagittarius
Mag.: —
Size: 20' by 20'

The Trifid Nebula is a challenging object because of its low surface brightness. The Trifid is the second in a bright pair of nebulae lying above the Teapot asterism of Sagittarius, the brighter of the two being the Lagoon Nebula (see #73). To bring yourself into the Lagoon-Trifid region, begin at Gamma Sagittarii and move 7° north. In a finder scope you'll see two distinct glows — the southernmost is the Lagoon and the northernmost the Trifid.

The Trifid Nebula is so named because of three prominent dark nebulae that float in front of the emission nebula and appear to join in the nebula's central region. These dark protrusions are visible in a 6-inch scope on a dark night. The emission nebula surrounds a bright triple star, GC 24537, which lies at the heart of the nebula. Just north of this region lies a second circular glow, much fainter than the primary nebula, that comes from bluish reflection gas surrounding a hot O-type star. The visibility of this nebulosity is greatly helped by using a nebula filter. (Incidentally, the bright open star cluster M21 lies 1° northeast of the Trifid Nebula.)

Photo by David Healy

73. The Lagoon Nebula (M8, NGC 6523)
Emission nebula
R.A. and Dec.: 18h04.7m -24°20'
Con.: Sagittarius
Mag.: —
Size: 90' by 40'

After you enjoy viewing the Trifid Nebula (see object #72), drop your telescope 2°, and you'll see the much larger and brighter emission nebula M8. This object consists of a broad fan of nebulosity stretching to the northeast and a brighter, oval patch of glowing gas to the west. These two main sections appear to be separated by a broad dust lane, giving the object its popular name, the Lagoon Nebula.

The Lagoon Nebula is the most prominent emission nebula in the summer Milky Way. It is a giant stellar birthplace, a region in which hot young stars are forming out of the recycled dust and gas from dead stars. Scattered throughout the northeastern part of the nebula are many white and blue-white suns that together constitute the star cluster NGC 6530, the group of stars forming from the gas clouds. This open cluster contains several dozen bright stars and adds greatly to the beauty of the area.

On dark nights large backyard instruments show a wealth of detail in the Lagoon Nebula. Aside from the two bright components, the region is blanketed in a soft, fuzzy glow that shows several sharply bright ridges running throughout. Averted vision helps to show the many small dark patches lying several arcminutes away from the central lagoon. This object is as grand an emission nebulosity as one can view from the northern hemisphere.

Photo by Kim Zussman

Photo by Martin C. Germano

74. The Parrot's Head Nebula
(Barnard 87)
Dark nebula
R.A. and Dec.: 18h04.3m -32°30'
Con.: Sagittarius
Mag.: —
Size: 12'

An unusually intriguing dark nebula can be found in central Sagittarius south of the Lagoon-Trifid region. B87, nicknamed the Parrot's Head Nebula, is a dark cloud opaque enough to be viewed with small telescopes and detailed enough to look interesting in astrophotos. To locate the rich area of stars containing B87, start at the bright star Gamma Sagittarii (the jumping-off point you used for finding the Lagoon and Trifid nebulae). Move 2° south of the bright star and be sure your scope is fitted with a low-power eyepiece.

The field containing B87, located not far from the galactic plane, is exceedingly rich in faint stars. This faint star bonanza is what helps to outline the form of the dark cloud that we see as B87. On a night when the transparency is good, a 4-inch scope will show an oval "hole" in the stars and what appears to be a small beak on one side. The central part of the nebula is partly filled in with faint stars. With averted vision you may see two thin lanes of dark material emanating outward from the bird's head. These appear to form wings visible only on the darkest night.

75. 70 Ophiuchi
Double star
R.A. and Dec.: 18h05.5m +2°30'
Con.: Ophiuchus
Mags.: 4.2, 6.0
Sep.: 2.5"

This pretty double star is composed of equal suns of magnitudes 4.2 and 6.0, whose total light equals that of a magnitude 4.0 star. To find 70 Ophiuchi, start by locating the bright naked-eye star Beta Ophiuchi. (You'll know you have Beta in sight when you see the bright, scattered open cluster IC 4665 2° northeast of the star.) From Beta, move 5° southeast to a wide triangle of 4th-magnitude stars. The easternmost of these stars is 70 Ophiuchi.

One of the earliest widely observed doubles, 70 Ophiuchi was discovered in 1779 by the English observer William Herschel. Its stars form a beautiful color contrast of yellow (primary) and orange or violet, and are currently separated by about 2.5". Because of the orbit of the secondary star, the components will widen to a separation of 6.7" in 2021. The real distance between the two stars is on average 2.1 billion miles, and the system lies approximately 16.5 light-years away. As such, it is one of the closest binary stars visible in small telescopes.

Photo by Robert Provin and Brad Wallis

Photo by David Healy

76. The Small Sagittarius Star Cloud
(M24)
Star cloud
R.A. and Dec.: 18h 15.0m -18°30'
Con.: Sagittarius
Mag.: —
Size: 120' by 40'

If you stand under the stars on a summer evening, one of the most striking features of the sky will greet you, the blazing center of the Milky Way. In the direction of Sagittarius you'll see light from innumerable star clouds and from the distant photons from tens of thousands of stars in the Sagittarius arm of our Galaxy. One of the most prominent naked-eye features of the Milky Way is the Small Sagittarius Star Cloud, a seemingly detached part of the Milky Way north of the Teapot along the galactic equator.

Charles Messier was so struck by this sizable (2° by 1°) bright patch that he gave it a number in his catalog of nebulous objects. Messier 24, as it is now known, is a gorgeous region for scanning with large binoculars or a telescope. Chief among the features associated with this star cloud are the small open cluster NGC 6603, a rich, 11th-magnitude object embedded in the cloud's glow and the easily visible dark nebulae B92 and B93 (see object #77).

Photo by Martin C. Germano

77. Barnard 92
Dark nebula
R.A. and Dec.: 18h15.5m -18°11'
Con.: Sagittarius
Mag.: —
Size: 12' by 6'

Lying on the northern edge of the Small Sagittarius Star Cloud (M24; see object #76) are the extremely opaque (and therefore easily visible) dark nebulae B92 and B93. Set against a background of innumerable faint stars, these clouds of fine particulate soot are closer to us than the stars and hence are made visible in outline by the stars lying beyond them.

B92 is the largest and most prominent of several dark nebulae in the central region of Sagittarius. Covering an area 12' by 6' across, M24 is impossible to miss when looking at it with a small telescope. You'll notice a single faint star lying near the center of the dark nebula — this is almost certainly a foreground object. Northeast of B92 lies the slightly less opaque but larger dark nebula B93, also easily within the visibility range of small telescopes. The overall shape of this nebula is that of a squashed triangle, with little lanes of stars seeping in around the edges.

Photo by Kim Zussman

78. The Eagle Nebula (M16, NGC 6611)
Emission nebula
R.A. and Dec.: 18h18.8m -13°47'
Con.: Serpens
Mag.: —
Size: 21'

One of many bright emission nebulae lying in the summer sky is M16, better known as the Eagle Nebula. This object is typical of large nebulae, beautifully detailed in long-exposure photos but because of low surface brightness challenging to visual observers. Nevertheless, with a 6-inch or larger scope, if the night is dark and transparent, you can see considerable detail in this distant stellar birthplace.

To locate M16, first find the bright star Alpha Scuti, which lies adjacent to the open cluster NGC 6664. Next, move south 7° and slightly west to an odd-shaped arrangement of 5th and 6th-magnitude stars arranged in a tight triangle and a close pair. From the brightest star in the pair, Gamma Scuti, move 3° west-northwest. That'll place you right on top of the Eagle Nebula.

This object's structure is complex. The most obvious feature in a small scope is the bright star cluster forming out of the M16 nebulosity. The nebula itself may require dark adaptation to see properly. When you see it, you'll note a broad, fan-shaped glow that has several prominent dark features superposed on it.

Photo by Kim Zussman

79. The Omega Nebula (M17, NGC 6618)
Emission nebula
R.A. and Dec.: 18h20.8m -16°11'
Con.: Sagittarius
Mag.: —
Size: 25'

A mere 3° south of the Eagle Nebula (see object #78), lying immediately south of a 6th-magnitude star, is another bright and wonderful emission nebula. M17, the Omega Nebula — also variously known as the Swan Nebula, the Horseshoe Nebula and the Checkmark Nebula — is far different in character from M16. Rather than containing a bright star cluster and possessing a low surface brightness — hallmarks of M16 — this nebula has a tremendously high surface brightness and offers almost no bright stars, just bright nebulosity.

Discovered by the Swiss astronomer Phillippe de Cheseaux in 1764, the Omega Nebula is one of the best for small telescopes. Its distinctive omega shape, formed by a bright bar of nebulosity and curving tendrils of glowing gas, is visible in a 2-inch telescope on a dark night. The 19th-century observer William H. Smyth called M17 "a magnificent arched and not resolvable luminosity occupying more than a third of the area in a splendid group of stars." Englishman William Herschel called it "a wonderful extensive nebulosity of the milky kind."

Whether you are using a 2-inch or 16-inch scope, M17 is one object guaranteed to provide a satisfying look into a large region of hydrogen gas.

80. M22 (NGC 6656)
Globular cluster
R.A. and Dec.: 18h36.5m -23°54'
Con.: Sagittarius
Mag.: 5.1
Size: 24'

Many observers claim that M13, the Hercules Cluster, is the most spectacular globular cluster visible from the northern hemisphere (see object #66). If any other cluster might give M13 a run for its money, that cluster is the brilliant M22 in Sagittarius.

M22 was discovered by the German astronomer Abraham Ihle in 1665. It can be found northeast of the constellation's Teapot asterism, about 3° east-northeast of Lambda Sagittarii. (The cluster is instantly recognizable as a fuzzy "star" between two clumps of 7th-magnitude stars aligned roughly northeast-southwest.)

Once M22 is located, its richness and easy resolution will astonish even the small telescope user. On a dark night a 6-inch scope reveals a large, milky glow 15' across that is alive with tiny, individual stellar images. The cluster's core is highly concentrated and difficult to resolve, but the edges are easily transformed into myriad stars and the face of the cluster can be completely resolved with an 8-inch scope. The group's halo of bright stars will leave any telescopist impressed.

81. R Scuti
RV Tauri-type variable star
R.A. and Dec.: 18h47.5m -5°42'
Con.: Scutum
Mag. range: 4.5-8.2
Period: 140 days

R Scuti is a semiregular variable star that is easy to follow with small telescopes. Discovered in 1795 by the Englishman E. Pigott, R Scuti was among the first dozen variable stars known. The star's spectral type ranges between G0 (maximum) and K0 (minimum), so it appears to fluctuate between yellow and orange as it brightens and dims. The magnitude range is 4.5 to 8.2, making it a naked-eye star during only part of its cycle. Normally, however, the star ranges between magnitudes 4.5 and 6.0. The drops in brightness below 6th magnitude come less frequently, usually every fourth or fifth minimum. The period is approximately 140 days.

This star is located in the rich Scutum Star Cloud, just 1° northwest of the open cluster M11 (see object #82) and 1° south of the bright star Beta Scuti.

Long mysterious, the mechanism for this star's variability is now known to be pulsation, the physical expansion and contraction of the star. R Scuti is thought to be about 100 times the diameter of the Sun, approximately 3,000 light-years distant, and at maximum shines with the luminosity of 8,000 Suns.

Photo by Jack Newton

Photo by Todd Owen

82. The Wild Duck Cluster
(M11, NGC 6705)
Open cluster
R.A. and Dec.: 18h51.1m -6°16'
Con.: Scutum
Mag.: 5.8
Size: 14'

The richest of all bright open clusters, M11 in Scutum is a beautiful sight for summer stargazers. Just visible to the naked eye as a fuzzy patch, this pretty cluster is composed of 500 stars of magnitudes 8 to 14, making it an unforgettable sight in telescopes.

M11 was first telescopically explored by the German astronomer Gottfried Kirch in 1681. Kirch described the group as "a small, obscure spot with a star shining through and rendering it more luminous." The star Kirch referred to is the cluster's brightest member, a centrally located orange star that shines at magnitude 8. The other stars are scattered in a triangle-shaped pattern that extends over 14'. Because of the bright apex star and flock of fainter suns that follow, 19th-century astronomers named this group the Wild Duck Cluster.

M11 is best viewed with a low-power eyepiece. An 8-inch scope at 50x provides a striking sight, showing the entire cluster and some surrounding stars.

Photo by Kim Zussman

83. The Ring Nebula (M57, NGC 6720)
Planetary nebula
R.A. and Dec.: 18h53.6m +33°02'
Con.: Lyra
Mag.: 8.8
Size: 86" by 62"

Aim your telescope at the correct spot in the tiny constellation Lyra, and you'll be rewarded by seeing a cosmic smoke ring. Although it measures just 86" by 62", the Ring Nebula has such a high surface brightness that it is easily visible in a 2-inch telescope.

To locate M57, guide your finder scope to the wide pair of stars Beta[1,2] and Gamma Lyrae, which help make up the southern end of the constellation. The Ring Nebula lies between these two bright stars, slightly closer to Beta[1,2] Lyrae.

Once you zoom in on the Ring, you'll be able to carefully view its structure. The ring shape and dark central hole are obvious in any instrument. Subtler features, such as the oblate shape of the ring, come with apertures of 6 inches or more. The real challenge in observing M57 lies in trying to spot the elusive, magnitude 14.8 central star. Some observers have reported seeing this faint point of light with scopes only 5 inches in diameter, while others have failed to see the star with 30-inch instruments. Seeing conditions and an observer's experience certainly play a part, but the star itself may be variable and therefore easier to see at certain times.

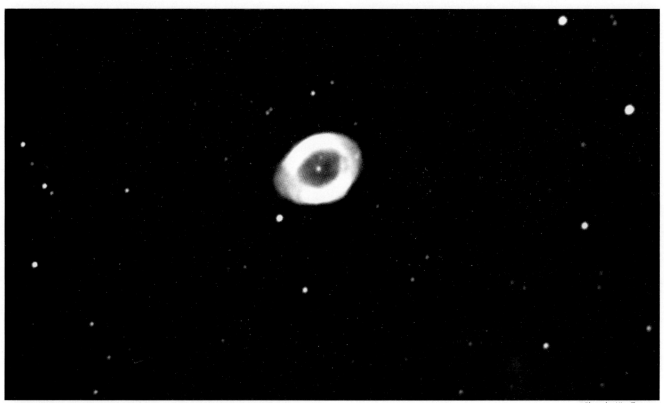

Photo by Kim Zussman

84. NGC 6781
Planetary nebula
R.A. and Dec.: 19h18.4m +6°33'
Con.: Aquila
Mag.: 11.4
Size: 1.9' by 1.8'

Another spherical planetary nebula lies in the summertime constellation Aquila. Much larger than the Ring Nebula (see object #83) but also fainter, NGC 6781 is trickier to see and warrants a moonless night.

Begin your journey to NGC 6781 by locating Delta Aquilae, the westernmost star that forms Aquila's distinctive V-shape. From Delta, move 3° to the 6th-magnitude star 22 Aquilae. Next, continue 1° northwest and you'll come upon a 6th-magnitude star. Now make a 90° turn and move northeast 1.5° to another 6th-magnitude star. NGC 6781 lies within the same low-power field as this star, immediately to the east.

NGC 6781 is a big bubble of nebulosity whose low surface brightness makes it a challenging object for small scopes. Nevertheless, a 6-inch instrument reveals a disk of light that is illuminated unevenly. The nebula lies in a rich star field and so appears like a ghostly apparition floating among bright pinpoints of light. A 10-inch scope reveals something of a three-dimensional effect with this nebula, as the uneven illumination of the sphere is more easily visible.

Photo by Martin C. Germano

85. Albireo (Beta Cygni)
Double star
R.A. and Dec.: 19h30.7m +27°58'
Con.: Cygnus
Mags.: 3.1, 5.1
Sep.: 34.4"

A bright, colorful double star in the summer sky is easy to find. Albireo is the bright naked-eye star that marks the nose of the swan in the constellation Cygnus. (It also marks the base of the large cross-shaped asterism that most people quickly recognize when viewing this group of stars.) Once viewed with a telescope, this ordinary star transforms into an extraordinary pair of stars.

Albireo's components glow with magnitudes of 3.1 and 5.1 and are dazzling in their colors, golden yellow and rich blue, respectively. The stars are separated by the enormous gap of 34.4", making them visible as a cleanly split pair in large

Photo by John Sanford

binoculars. The distance to this binary system is about 400 light-years, making the gap between the two stars about 400 billion miles. Curiously, astronomers have observed a composite spectrum from the primary star, suggesting it consists of two stars too close to see as individuals.

86. The Blinking Planetary (NGC 6826)
Planetary nebula
R.A. and Dec.: 19h44.8m +50°31'
Con.: Cygnus
Mag.: 8.8
Size: 27" by 24"

Although small and appearing nearly starlike at low power, the planetary nebula NGC 6826 in Cygnus displays a curious effect. When you look at the nebula directly, the disk of nebulosity is invisible, leaving only a magnitude 10.4 central star. When you glance at the corner of your eyepiece field, employing averted vision, the sensitive rods in your eyes pick up the nebulous disk and you see the nebula itself. If you observe NGC 6826 and alternate between a direct view and a side glance, the nebula appears to blink on and off.

This peculiar object can be found in the northwestern reaches of Cygnus. You can star-hop there by starting with Delta Cygni and moving 7° north-northwest to Iota Cygni. Next, move 2° southeast to the 6th-magnitude double star 16 Cygni. From there, simply swing 1° east and you'll be aimed exactly toward NGC 6826.

The so-called Blinking Planetary is especially noticeable because of its bluish color. Once you've located the object, switch to high magnification and take a close look at the Blinking effect. It's the result of a strange combination of eye, telescope, and photons that will leave you remembering NGC 6826 as a special object.

87. M71 (NGC 6838)
Globular cluster
R.A. and Dec.: 19h53.7m +18°47'
Con.: Sagitta
Mag.: 8.0
Size: 7.2'

Prominent on the list of autumn deep-sky showpieces is the rich globular cluster M71, an object located in the small, arrow-shaped constellation Sagitta. M71 is so loose and easily resolved by small telescopes that for decades after its discovery astronomers thought it might be a rich open cluster. Not so — it is a bonafide globular that is a gorgeous sight on cool autumn evenings.

To find M71, start with the bright star Delta Sagittae, the central star in the constellation's arrow-shaped asterism. Move 2° east and you'll swing by a bright double star and then M71, immediately east of the star. After locating the cluster, be sure you have a moderately high magnification eyepiece in place (about 150x) for a close look at the group.

At this magnification, you'll see the brightest stars in the cluster — especially those on the group's edges — as individual points. Overall, the shape of M71 is somewhat triangular, and it lies in an exceedingly rich star field that appears as a glistening backdrop. On ultra-dark nights you may be able to resolve the stars of M71 clean across its face, yielding a stunning view of a great, distant globe of aged suns.

Photo by Jack Newton

Photo by Martin C. Germano

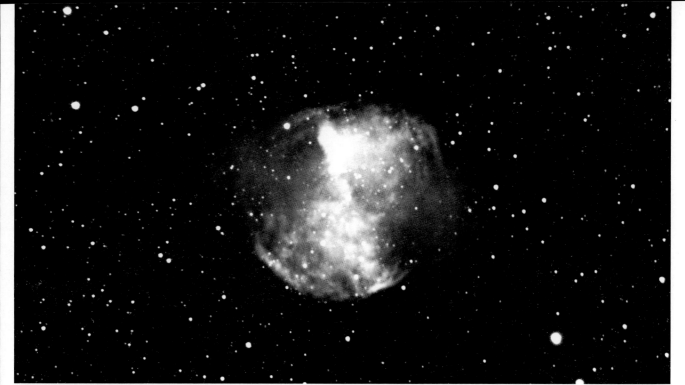

Photo by Kim Zussman

88. The Dumbbell Nebula
(M27, NGC 6853)
Planetary nebula
R.A. and Dec.: 19h59.6m +22°43'
Con.: Vulpecula
Mag.: 7.3
Size: 8.0' by 5.7'

Tucked away in the diminutive constellation Vulpecula is a grand planetary nebula that will impress you in any telescope. The Dumbbell Nebula, so-named because of its two-lobed structure, is enormous and very bright as planetaries go. Measuring 8' by 5.7' and glowing at 7th magnitude, the Dumbbell is visible in binoculars and small finder scopes as a hazy smudge of gray light. Large instruments reveal a wealth of detail in this object, including its central star and "ears" of faint nebulosity extending far out from the bright center.

The best star-hopping routine for M27 begins at the bright star Albireo (Beta Cygni, see object #85). From Albireo move 7° east-southeast to a bright pair of stars, one of which is the 4th-magnitude double 13 Vulpeculae. Continue along this path for another 2°, and you'll see a double glow in the finder, the 6th-magnitude star 14 Vulpeculae and a hazy, circular glow, the Dumbbell.

A 4-inch scope shows M27 as a double glow set in a dazzlingly starry field. A 6-incher provides a better view in that the dumbbell shape is clearly visible (some observers liken this form to a butterfly or an apple core). A good 8-inch glass shows a double wing structure, the 14th-magnitude central star, and faint outer nebulosity on both sides of the nebula's bright parts. The surface brightness of M27 is high, so don't be afraid to experiment with high-power viewing on this object.

89. NGC 6888
Supernova remnant
R.A. and Dec.: 20h12.5m +38°25'
Con.: Cygnus
Mag.: —
Size: 20' by 10'

Riding high on the galactic plane in central Cygnus is a little-known but beautiful supernova remnant called NGC 6888, occasionally known as the Crescent Nebula. This object appears quite spectacular and detailed in photographs but its low surface brightness makes it a challenge for small telescope users. Nevertheless, NGC 6888 is visible given the right conditions and observing techniques.

Locating the position of NGC 6888 is simple. Start at the 3rd-magnitude double star Sadr (Gamma Cygni), the central star in the "cross" of Cygnus. Move southwest by 2.5° and you'll come to a tight, arrow-shaped grouping of 7th and 8th-magnitude stars. Next, move 1° west and you'll be centered on the field of NGC 6888.

There are several ways to spot this tricky object. The best beginning is to use a low-power eyepiece. Try simply scanning the field, looking for faint nebulosity in the corner of your eye by averted vision. Be sure you're observing on a night that is moonless. Try using a red-transmissive nebula filter to maximize the contrast between the sky and faint light from the object. On a good dark night, you'll see a ghostly shell of light some 20' by 10' in extent. This light represents the expanding remnant of a star that exploded some 20,000 years ago.

Photo by Alfred Lilge

90. The Veil Nebula
Supernova remnant

NGC 6960
R.A. and Dec.: 20h45.7m +30°43'
Con.: Cygnus
Mag.: —
Size: 70' by 6'

NGC 6979
R.A. and Dec.: 20h48.5m +31°09'
Con.: Cygnus
Mag.: —
Size: 45' by 30'

NGC 6992
R.A. and Dec.: 20h56.4m +31°42'
Con.: Cygnus
Mag.: —
Size: 60' by 8'

NGC 6995
R.A. and Dec.: 20h57.1m +31°13'
Con.:Cygnus
Mag.: —
Size: 12'

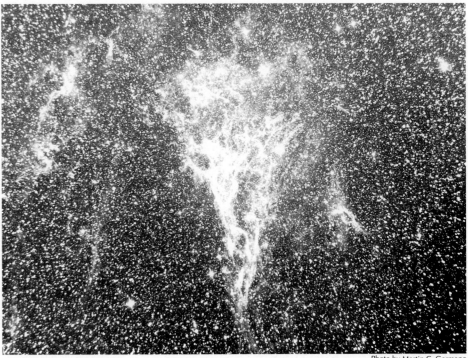

Photo by Martin C. Germano

The Veil Nebula in Cygnus is a supernova remnant that is far larger and brighter than NGC 6888 (see object #89), and is correspondingly much easier to observe. Because it consists of a network of separate filaments spread over several degrees of sky, the best method of observing the Veil Nebula is one piece at a time.

To find the best starting place, all you need to do is locate the star 52 Cygni. From Epsilon Cygni, move 4° south and you'll be pointing directly at 52, a 4th-magnitude double star. NGC 6960, one of the two main segments of the Veil Nebula, lies in back of this star and appears to pass through it as it winds through a north-south length of 70'. The other main parts of this nebula, NGC 6992 and NGC 6995, lie approximately 3° east of NGC 6960. This north-south segment winds through a rich star field and is considerably brighter and more detailed than NGC 6960. A more challenging piece of the shell is NGC 6979. This triangular-shaped object lies about 1.5° northeast of 52 Cygni.

In telescopes, the Veil Nebula is pure enjoyment. Both NGC 6960 and NGC 6992-5 show intricate structure in telescopes as small as 6 inches in aperture. The characteristic features of these nebulae are small, bright ropy filaments that twist their way through a background glow of smoother light. NGC 6979 is considerably fainter, but its triangular shape is visible in 10-inch and larger scopes under a dark sky.

Photo by Martin C. Germano

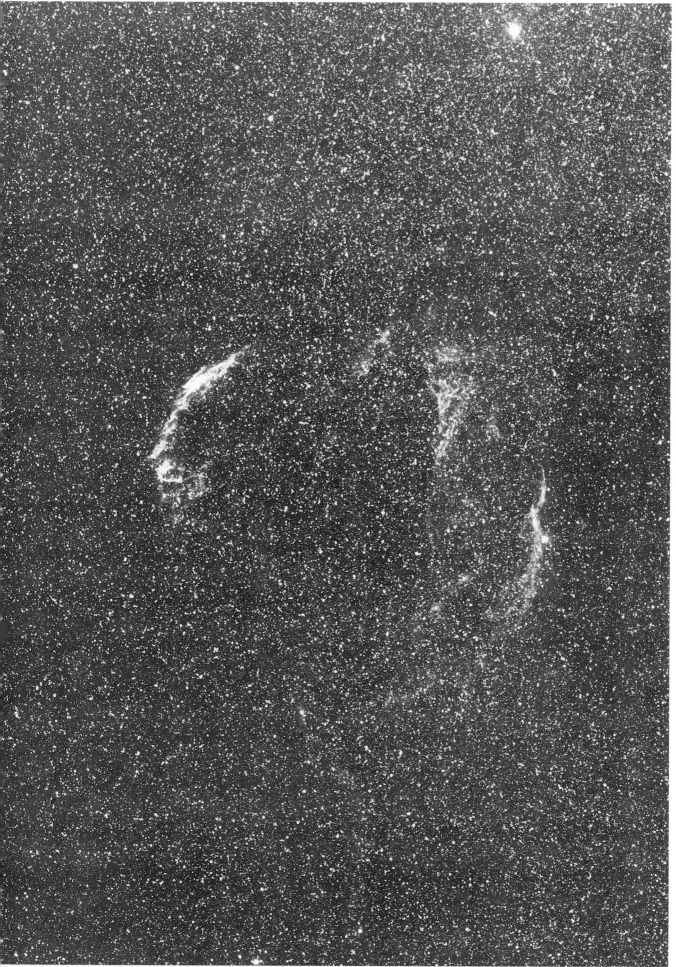

Photo by Alfred Lilge

91. The North America Nebula
(NGC 7000)
Emission nebula
R.A. and Dec.: 20h58.8m +44°20'
Con.: Cygnus
Mag.: —
Size: 100' by 60'

Visible to the naked eye when Cygnus hangs overhead, the North America Nebula is rarely observed but frequently photographed. Legendary for its appearance in long-exposure photos (in which it resembles North America), the nebula is less often linked with visual observing. Part of this is because the North America Nebula is so large — too large for telescopic fields and many binoculars. Yet the North America Nebula is much easier to observe than most backyard astronomers believe.

The best way to observe the North America Nebula is to venture out on a summer night when Cygnus is near the zenith. After fully dark-adapting your eyes, look straight up toward Deneb (Alpha Cygni). You should see a large, diffuse glow — something that resembles a bright portion of the Milky Way — immediately southeast of Deneb. This glow comes from the North America Nebula (and the fainter nebula west of the North America, IC 5067-70, known as the Pelican Nebula). To see the shape of the nebula better defined, try holding a red transmissive nebula filter right up to your eye as you look skyward. This should help you see the outline resembling the shape of North America.

In wide-field binoculars the North America Nebula is an unforgettable sight, studded with rich groups of bright stars. (The loose open cluster NGC 6997 lies in the northeastern part of the nebula.) The brightest parts of NGC 7000 are Mexico, the Midwest, and the Eastern Seaboard.

92. M15 (NGC 7078)
Globular cluster
R.A. and Dec.: 21h30.0m +11°10'
Con.: Pegasus
Mag.: 6.0
Size: 12'

M15 is a striking, condensed globular cluster best visible in the autumn sky, a time when relatively few globulars can be seen. At 6th-magnitude, this cluster is just visible to the naked eye, but it is really worthwhile observing with a telescope 6 inches or more in aperture.

Find M15 by starting at the bright star Enif (Epsilon Pegasi), the 2nd-magnitude double star that outshines everything in its region. From Enif, move 4° northwest to a pair of 6th-magnitude points of light separated by less than 1°. The westernmost "star" of the two is M15.

With a 4-inch scope at 100x, M15 appears as a bright core of hazy light surrounded by a few resolved stellar points. An 8-incher nicely resolves this cluster into hundreds of diamond-like stars. Large backyard telescopes reveal a condensed, ball-like center and weak chains of stars that wind away from the core much like the brighter cluster M13 (see object #66).

Photo by Fred Espenak

93. M39 (NGC 7092)
Open cluster
R.A. and Dec.: 21h32.2m +48°26'
Con.: Cygnus
Mag.: 4.6
Size: 32'

The sprawling, loose open star cluster M39 is a perfect object for binocular stargazers scanning the summer Milky Way. Visible as a hazy patch to the naked eye, this group covers the same area as the Full Moon and is so loose it appears best in binoculars or finder scopes rather than telescopes. Finding M39 is a snap. Start by centering your finder or binoculars on Deneb (Alpha Cygni), the brightest star in Cygnus. Move 8° east to Rho Cygni, a 4th-magnitude star lying in a rich Milky Way field. Next, move 3.5° north and the bright glow of M39 will come into view.

The brightest stars in M39 shine at 7th magnitude, and altogether 30 bright stars make up this cluster's population. This cluster was probably observed as early as 325 B.C., when Aristotle recorded what is believed to be M39 as a "cometary object." In the 19th century the English observer William H. Smyth called M39 simply a "splashy field of stars."

Photo by Lee C. Coombs

Photo by Jack Newton

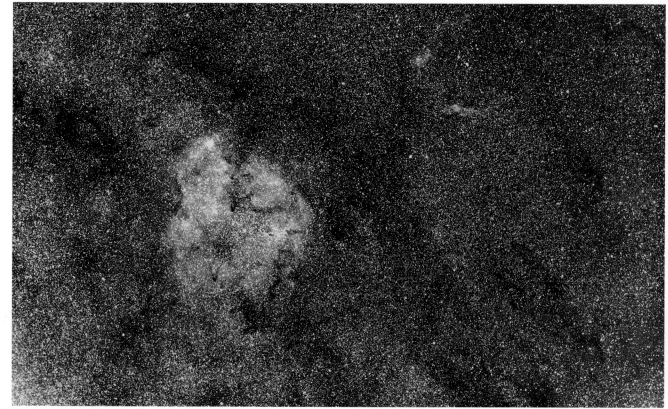

Photo by Alfred Lilge

94. IC 1396
Emission nebula
R.A. and Dec.: 21h39.1m +57°28'
Con.: Cepheus
Mag.: —
Size: 170' by 130'

IC 1396 is an emission nebula located relatively nearby that, despite its visibility to the naked eye, is almost totally ignored by amateur astronomers. Perhaps from being overshadowed by the bright sights of Cygnus, perhaps because it is too large to be viewed telescopically, or perhaps because people simply don't know it's visible, few amateurs turn their gaze toward this magnificent object. Yet it is visible as a bright circular glow in the Milky Way and is a detailed nebula full of dark lanes in backyard astrophotographs.

This nebula lies in a star-studded region several degrees away from the galactic equator in the constellation Ce- pheus. The easiest way to find IC 1396 is to locate Mu Cephei, a ruddy, 4th-magnitude variable known as Herschel's Garnet Star. IC 1396 consists of a scattered open cluster and large wreath-shaped glow that covers 3° by 2.5° just south of the Garnet Star.

A red-transmissive nebula filter helps significantly with spotting this nebula by eye alone, and indeed a dark sky is required to separate it from the general glow of the Milky Way. Binoculars provide the best view of IC 1396, permitting you to peer into a region of hundreds of faint stars and the large, hazy glow of hydrogen gas.

95. SS Cygni
U Geminorum-type variable star
R.A. and Dec.: 21h42.7m +43°35'
Con.: Cygnus
Mag. range: 8.2-12.4
Period: 50 days

SS Cygni is a well-observed variable star that was discovered at Harvard College Observatory in 1896. It is a so-called cataclysmic variable or dwarf nova, in which the star normally glows faintly at 12th magnitude but several times per year explodes into brightness. During these maxima, the star can rise four magnitudes in brightness, usually peaking at magnitude 8. These outbursts normally take place in 24 to 48 hours, and the average interval between the explosions is about 50 days.

SS Cygni is easily located in the region of Cygnus east of the bright star Deneb (Alpha Cygni). From Deneb, move 8° east to Rho Cygni, a 4th-magnitude star. Next, move your finder scope 3° south-southeast to the double star 75 Cygni. SS Cygni lies less than 1° northeast of this star.

Spectroscopic studies show that SS Cygni is a close double star consisting of a hot blue star and a yellow dwarf separated by less than 100,000 miles — nearly in contact on a stellar scale. The nova-like outbursts of the system are most likely caused by interaction between the two stars.

Photo by Todd Owen

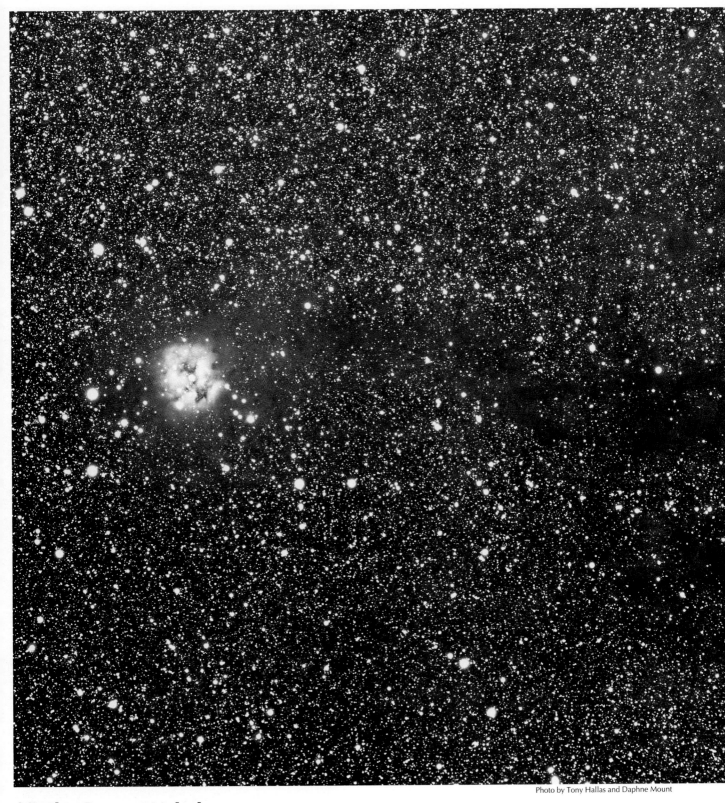

Photo by Tony Hallas and Daphne Mount

96. The Cocoon Nebula (IC 5146)
Emission nebula
R.A. and Dec.: 21h53.2m +47°16'
Con.: Cygnus
Mag.: —
Size: 10' by 10'

High up in the thick summer Milky Way in the northern reaches of Cygnus lies a peculiar, dust-enshrouded emission nebula designated IC 5146. Because it appears as a circular, glowing sphere and contains so much dust, astronomers have nicknamed this object the Cocoon Nebula. It is a challenge to find because it is so dim that it has little contrast against the sky.

Begin your journey to the Cocoon Nebula by locating Deneb (Alpha Cygni), the brightest star in the constellation. Move 8° east to the 4th-magnitude gem Rho Cygni, then 4° northeast to the equally-bright star Pi^2 Cygni. The Cocoon Nebula lies 3° south-southeast of Pi^2 Cygni, at the end of a long, winding lane of dark nebulosity.

The visibility of the Cocoon is greatly dependent on sky conditions. This is an object that should be reserved for the darkest, most transparent nights, and then the proper instrument is a wide-field telescope at least 6 inches in aperture. With such a scope the nebula appears as a soft glow about a third the size of the Full Moon with a bright central star and several dark patches crossing its ghostly disk.

97. The Helical Nebula (NGC 7293)
Planetary nebula
R.A. and Dec.: 22h29.6m -21°51'
Con.: Aquarius
Mag.: 7.3
Size: 12' by 10'

The Helical Nebula is a giant contradiction. Easily the closest, largest, and brightest planetary nebula in the sky, it is also one of the most challenging to observe. The culprit is the object's large total size, which spreads its brightness over a large area. The result is low surface brightness. In photographs, the nebula shows the double helical spiral pattern that provided its nickname. Visually, however, the nebula is a tricky customer, compounded by the fact that it lies low in the sky for most northern hemisphere viewers.

The Helix lies in a star-poor region of Aquarius, a relatively dim autumnal constellation. Start your star-hop to this nebula by finding Fomalhaut (Alpha Piscis Austrinus), the brightest star in the present region of sky. Move 4° west-northwest to 4th-magnitude Epsilon Piscis Austrinus, then 7° north-northwest to 5th-magnitude Nu Aquarii. About 3.5° southwest of Nu lies the similarly-bright star 47 Aquarii. The Helical Nebula lies between 47 and Nu Aquarii, somewhat closer to Nu.

In the eyepiece you'll see the nebula as a large, circular haze. A good 8-inch scope reveals the dark, central "hole" in the object and some detail in the ring structure of the nebula. The object is best viewed with a large backyard scope equipped with a red-transmissive nebula filter under pitch-black skies.

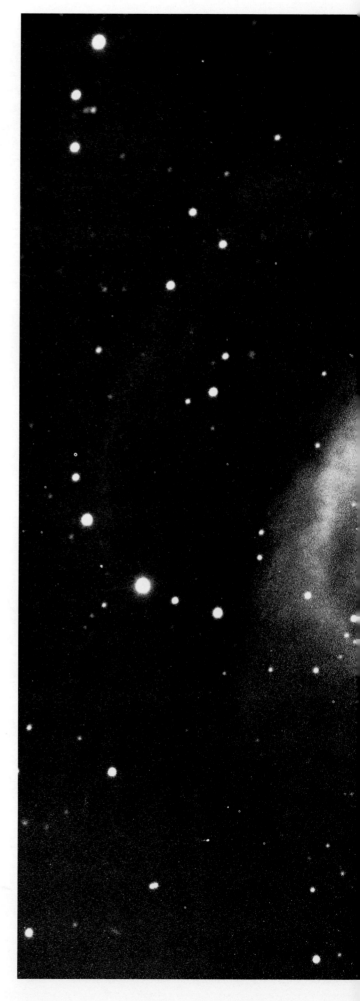

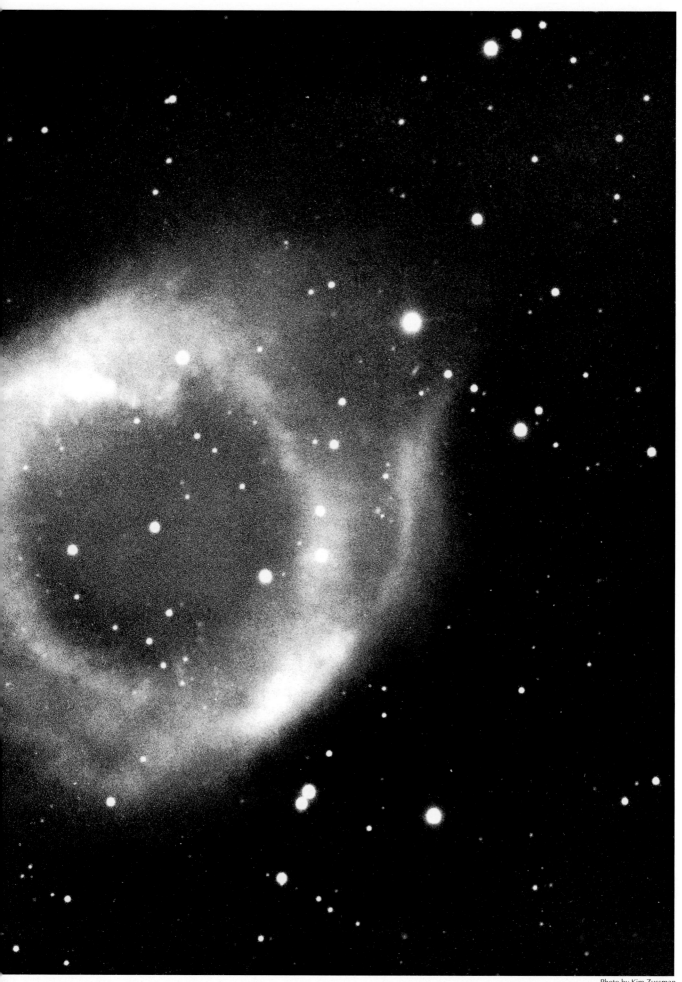

Photo by Kim Zussman

98. NGC 7331
Spiral galaxy
R.A. and Dec.: 22h37.1m +34°26'
Con.: Pegasus
Mag.: 9.5
Size: 11' by 4.0'

Pegasus conceals from the naked eye an alluring spiral galaxy that is prominently visible in backyard scopes. NGC 7331 is a textbook Sb-type spiral, much like our Milky Way, that lies some 50 million light-years away. Curiously, it is structurally similar to the Andromeda Galaxy, the closest spiral to us (see object #2). Yet NGC 7331 provides us with a compelling picture of what the Andromeda Galaxy might look like if it were located 25 times farther away.

NGC 7331 offers an easy star-hop. First find Eta Pegasi, the 2nd-magnitude star that forms part of the equilateral triangle in the northwestern corner of the Great Square of Pegasus. Now move 4.5° north-northwest and you'll be placed directly on the NGC 7331 field of view.

In small telescopes NGC 7331 shows a bright, oval nucleus surrounded by a nebulous glow some 9' by 3' in extent. Larger telescopes — those in the 16-inch range — reveal knotty, dark structure in the galaxy's arms when viewing at high power. As a telescopic side trip, move 30' south-southwest and you'll find a small group of five faint galaxies. These objects — NGC 7317, NGC 7318A, NGC 7318B, NGC 7319, and NGC 7320 — compose the galaxy group called Stephen's Quintet.

Photo by Kim Zussman

99. The Bubble Nebula (NGC 7635)
Emission nebula
R.A. and Dec.: 23h20.7m +60°10'
Con.: Cassiopeia
Mag.: —
Size: 15' by 8'

The Bubble Nebula is an emission region in which hot stars have blown away a shell, or bubble, of gas. This object is a low surface brightness nebula that requires an exceptionally dark night at a telescope of 6 inches or more aperture to see well.

Start your star-hop to the Bubble Nebula by finding Beta Cassiopeiae, the westernmost star in the constellation's familiar W-shaped pattern. From Beta, move 5.5° northwest to the 5th-magnitude star 4 Cassiopeiae. The Bubble Nebula lies 1.5° south of this star. (Two other deep-sky objects are within easy range: open cluster M52 is less than 1° northeast of the Bubble Nebula, and emission nebula NGC 7538 is 1° west of the Bubble.)

With an 8-inch scope at 100x, the Bubble Nebula appears as a wisp of milky light floating in a rich field of faint stars. Larger scopes show the object much better, and the spherical bubble itself is clearly visible in a 16-inch instrument. This object is definitely one to reserve for the darkest, clearest nights.

Photo by Preston S. Justis

Photo by Preston S. Justis

100. NGC 7789
Open cluster
R.A. and Dec.: 23h57.0 +56°44'
Con.: Cassiopeia
Mag.: 6.7
Size: 16'

A resident of the northern reaches of Cassiopeia, star cluster NGC 7789 is an exceedingly rich cluster containing faint stars. As such, it is visible in small scopes as a faint cloud of light and is cleanly resolved into pinpoint stars only in telescopes of 6 inches and more aperture.

NGC 7789 lies in a field peppered with faint stars. Find the cluster by star-hopping from bright star Beta Cassiopeiae. Slightly over 3° southwest of Beta lies a wide pair of stars consisting of Rho and Sigma Cassiopeiae. These stars are aligned north-south and separated by just over 2°. NGC 7789 lies right smack between these two stars.

A 6-inch telescope shows a cloud of more than 200 stars covering an area 16' across. Larger telescopes, with their narrower fields of view, simply do not do justice to NGC 7789. This is one object specifically designed for the small telescope astronomer.

Photo Credits

1. 47 Tucanae: Harvard College Observatory.

2. M31, M32, and NGC 205: Tony Hallas and Daphne Mount, 5-inch f/8 Starfire refractor, hypersensitized 6415 Tech Pan film.

3. NGC 253: David Healy, 14-inch SCT at f/7, hypersensitized Tech Pan film, 40-minute exposure.

4. The Magellanic Clouds: Ronald Royer.

4A. The Large Magellanic Cloud: Kitt Peak National Observatory. 4B. The Small Magellanic Cloud: Harvard College Observatory.

5. Globular Cluster NGC 362: Cerro Tololo Interamerican Observatory

6. The Owl Cluster: Lee C. Coombs, 10-inch f/5 reflector, 103a-O film, 5-minute exposure.

7. M33: Kim Zussman, modified 11-inch f/10 SCT, hypersensitized Tech Pan film, 120-minute exposure.

8. The Little Dumbbell Nebula: Kim Zussman, modified 11-inch f/10 SCT, hypersensitized Tech Pan film, 90-minute exposure.

9. Gamma Andromedae: Ron Miller, 8-inch f/10 SCT, 1600 ISO film, 15-second exposure.

10. The Double Cluster: Martin C. Germano, 8-inch f/6 reflector, hypersensitized Tech Pan film, 30-minute exposure.

11. Mira: Robert Provin.

12. Algol: Robert Provin and Brad Wallis, 35mm f/4 lens, hypersensitized Fujichrome 100 film, 50-minute exposure.

13. NGC 1360: Martin C. Germano, 8-inch f/5 reflector, 103a-F film, 25-minute exposure.

14. The Pleiades: Rick Dilsizian, 8-inch f/1.5 Schmidt camera, hypersensitized Tech Pan film, 7-minute exposure.

15. The California Nebula: Alfred Lilge, 4-inch f/6 astrograph, hypersensitized Tech Pan film, 120-minute exposure.

16. The Hyades: J.A. Farrell, 7.5-inch Schmidt camera, Royal Pan film, 5-minute exposure.

17. The Crab Nebula: Neyle Sollee.

18. The Orion Nebula: Tony Hallas and Daphne Mount.

19. The Tarantula Nebula: Steve Quirk.

20. The Horsehead Nebula: Preston S. Justis, 10-inch f/6 reflector, 103a-F film, 86-minute exposure.

21. M37: Martin C. Germano, 8-inch f/10 SCT, 103a-F film, 40-minute exposure.

22. M35: Martin C. Germano, 8-inch f/5 reflector, 103a-F film, 15-minute exposure.

23. The Rosette Nebula: Tony Hallas and Daphne Mount, 6-inch EDF refractor, hypersensitized 6415 Tech Pan film, 120-minute exposure.

24. Hubble's Variable Nebula: Paul Roques, 16-inch f/4.5 reflector, hypersensitized Tech Pan film, 35-minute exposure.

25. The Cone Nebula: Martin C. Germano, 8-inch f/5 reflector, hypersensitized Tech Pan film, 45-minute exposure.

26. Sirius: Jim Barclay, 12.5-inch f/6 reflector, Fujichrome 100 film, 2-minute exposure.

27. M41: Martin C. Germano, 8-inch f/5 reflector, 103a-F film, 15-minute exposure.

28. The Eskimo Nebula: Jack Newton, 20-inch reflector at f/10, hypersensitized Tech Pan film.

29. NGC 2403: Martin C. Germano, 8-inch f/10 SCT, 103a-F film, 70-minute exposure.

30. M46 and NGC 2438: Bill Iburg, 8-inch SCT at f/5, 103a-F film, 35-minute exposure.

31. NGC 2516: 200mm f/2.5 lens, Fujichrome 400 film, 7-minute exposure.

32. Zeta Cancri: Robert Provin and Brad Wallis.

33. The Beehive Cluster: Jay Anderson, 8-inch f/1.5 Schmidt camera, hypersensitized Tech Pan film, 16-minute exposure.

34. NGC 2841: Martin C. Germano, 8-inch f/10 SCT, 103a-F film, 90-minute exposure.

35. NGC 2903: Paul Roques, 16-inch f/4.5 reflector, hypersensitized Tech Pan film, 60-minute exposure.

36. M81: Kim Zussman, modified 11-inch f/10 SCT, hypersensitized Tech Pan film, 120-minute exposure.

37. M82: Kim Zussman, modified 11-inch f/10 SCT, hypersensitized Tech Pan film, 50-minute exposure.

38. The Ghost of Jupiter: Jack Marling, 24-inch f/15 Cassegrain, Fujichrome 400 film, 10-minute exposure.

39. IC 2602: Gregg D. Thompson, 6-inch reflector, 15-minute exposure.

40. The Eta Carinae Nebula: Allan Green, 5.5-inch f/2 Schmidt camera, hypersensitized Tech Pan film, 12-minute exposure.

41. The Owl Nebula: Kim Zussman, modified 11-inch f/10 SCT, hypersensitized Tech Pan film, 120-minute exposure.

42. M65 and M66: K. Alexander Brownlee, 16-inch f/5 reflector.

43. M99: Lee C. Coombs, 10-inch f/5 reflector, 103a-O film, 15-minute exposure.

44. M106: Kim Zussman, modified 11-inch f/10 SCT, hypersensitized Tech Pan film, 120-minute exposure.

45. M100: Martin C. Germano, 8-inch f/5 reflector, hypersensitized Tech Pan film, 35-minute exposure.

46. The Coma Star Cluster: Alan Dyer, 300mm f/5.6 lens, Tri-X film, 10-minute exposure.

47. NGC 4565: Kim Zussman, modified 11-inch f/10 SCT, hypersensitized Tech Pan film, 150-minute exposure.

48. The Sombrero Galaxy: Kim Zussman, modified 11-inch f/10 SCT, hypersensitized Tech Pan film, 90-minute exposure.

49. Porrima: John Sanford, 24mm f/3 lens, Fujichrome R100 film, 11-minute exposure.

50. M94: Martin C. Germano, 8-inch f/10 SCT, 103a-F film, 60-minute exposure.

51. The Coal Sack: Ronald Royer, 4-inch f/5 astrograph, tricolor negative from three exposures.

52. The Jewel Box Cluster: Chris Floyd, 10-inch reflector.

53. The Blackeye Galaxy: Kim Zussman, modified 11-inch f/10 SCT, hypersensitized Tech Pan film, 95-minute exposure.

54. M63: Paul Roques, 10-inch f/10 SCT, hypersensitized Tech Pan film, 104-minute exposure.

55. Mizar: Robert Provin and Brad Wallis.

56. Centaurus A: Jim Barclay, 8-inch f/6 reflector, Ilford HP 5 film, 30-minute exposure.

57. Omega Centauri: Jean Dragesco, 300mm f/1.5 Schmidt camera, hypersensitized Tech Pan film, 5-minute exposure.

58. The Whirlpool Galaxy: Kim Zussman, modified 11-inch f/10 SCT, hypersensitized Tech Pan film, 115-minute exposure.

59. M83: Martin C. Germano, 8-inch f/6 reflector, hypersensitized Tech Pan film, 60-minute exposure.

60. M3: Martin C. Germano, 8-inch f/10 SCT, 103a-F film, 35-minute exposure.

61. M101: Kim Zussman, modified 11-inch f/10 SCT, hypersensitized Tech Pan film, 120-minute exposure.

62. Xi Bootis: Daniel J. McGlaun Jr., 50mm f/2 lens, Ektachrome 400 film, 2-minute exposure.

63. M5: Lee C. Coombs, 10-inch f/5 reflector, 103a-O film, 15-minute exposure.

64. R Coronae Borealis: Orien Ernest.

65. M4: Alfred Lilge, 12.5-inch f/8 Proscope, 098-04 film, 120-minute exposure.

66. The Hercules Cluster: Rick Dilsizian, 14-inch SCT at f/7, hypersensitized Tech Pan film, 27-minute exposure.

67. NGC 6231: Martin C. Germano, 8-inch f/5 reflector, hypersensitized Tech Pan film, 20-minute exposure.

68. M92: Lee C. Coombs, 10-inch f/5 reflector, 103a-O film, 10-minute exposure.

69. The S Nebula: Martin C. Germano, 8-inch f/6 reflector, hypersensitized Tech Pan film, 65-minute exposure.

70. The Butterfly Cluster: Lee C. Coombs, 10-inch f/5 reflector, 103a-O film, 5-minute exposure.

71. M7: Martin C. Germano, 8-inch f/5 reflector, 103a-F film, 15-minute exposure.

72. The Trifid Nebula: Kim Zussman, modified 11-inch f/10 SCT, hypersensitized Tech Pan film, 120-minute exposure.

73. The Lagoon Nebula: David Healy, 14-inch SCT at f/7, hypersensitized Tech Pan film, 24-minute exposure.

74. The Parrot's Head Nebula: Martin C. Germano, 8-inch f/5 reflector, 103a-F film, 20-minute exposure.

75. 70 Ophiuchi: Robert Provin and Brad Wallis, 50mm f/4.5 lens, GAF 200 film, 40-minute exposure.

76. The Small Sagittarius Star Cloud: David Healy, 8-inch f/1.5 Schmidt camera, Tech Pan film, 15-minute exposure.

77. Barnard 92: Martin C. Germano, 8-inch f/5 reflector, 103a-F film, 25-minute exposure.

78: The Eagle Nebula: Kim Zussman, modified 11-inch f/10 SCT, hypersensitized Tech Pan film, 70-minute exposure.

79: The Omega Nebula: Kim Zussman, modified 11-inch f/10 SCT, hypersensitized Tech Pan film, 30-minute exposure.

80. M22: Jack Newton, 12.5-inch f/5 reflector, 103a-E film, 10-minute exposure.

81. R Scuti: Todd Owen.

82. The Wild Duck Cluster: Kim Zussman, modified 11-inch f/10 SCT, hypersensitized Tech Pan film, 15-minute exposure.

83. The Ring Nebula: Kim Zussman, modified 11-inch f/10 SCT, hypersensitized Tech Pan film, 20-minute exposure.

84. NGC 6781: Martin C. Germano, 8-inch f/10 SCT, 103a-F film, 40-minute exposure.

85. Albireo: John Sanford, 8-inch f/10 SCT, KB-14 film, 3-second exposure.

86. The Blinking Planetary: Jack Newton, 12.5-inch f/6 reflector, Ektachrome 400 film, 20-minute exposure.

87. M71: Martin C. Germano, 8-inch f/10 SCT, 103a-F film, 70-minute exposure.

88. The Dumbbell Nebula: Kim Zussman, modified 11-inch f/10 SCT, hypersensitized Tech Pan film, 70-minute exposure.

89. NGC 6888: Alfred Lilge, 12.5-inch f/8 Proscope, 098-04 film, 120-minute exposure.

90. The Veil Nebula: Alfred Lilge, 7-inch f/3.5 astrograph, 098-04 film, 25-minute exposure with a red filter.

90A. NGC 6992-5: Martin C. Germano, 8-inch f/5 reflector, hypersensitized Tech Pan film, 40-minute exposure.

90B. NGC 6960: Martin C. Germano, 8-inch f/10 SCT, 103a-F film, 90-minute exposure.

90C. NGC 6979: Martin C. Germano, 8-inch f/5 reflector, hypersensitized Tech Pan film, 65-minute exposure.

91. The North America Nebula: Jack Newton, 300mm f/2.8 lens, hypersensitized Tech Pan film, 30-minute exposure with a red filter.

92. M15: Fred Espenak, 8-inch SCT at f/7, hypersensitized Tech Pan film, 25-minute exposure.

93. M39: Lee C. Coombs, 10-inch f/5 reflector, 103a-O film, 10-minute exposure.

94. IC 1396: Alfred Lilge, 7-inch f/3.5 astrograph, 098-04 film, 45-minute exposure with a red filter.

95. SS Cygni: Todd Owen.

96. The Cocoon Nebula: Tony Hallas and Daphne Mount, 5-inch f/8 Starfire refractor, hypersensitized 6415 Tech Pan film, 120-minute exposure.

97. The Helical Nebula: Kim Zussman, modified 11-inch f/10 SCT, hypersensitized Tech Pan film, 120-minute exposure.

98. NGC 7331: Kim Zussman, modified 11-inch f/10 SCT, hypersensitized Tech Pan film, 160-minute exposure.

99. The Bubble Nebula: Preston S. Justis, 10-inch f/6 reflector, 103a-F film, 80-minute exposure.

100. NGC 7789: Preston S. Justis, 10-inch f/6 reflector, 103a-O film, 65-minute exposure.

Index

A

Acrux, 23
Albireo, 64, 65
Alcor, 51
Alcyone, 12
Aldebaran, 10, 13
Algol, 4, 11
Almach, 9
Almagest, 55
Alpha Bootis, 51
Alpha Canis Majoris, 19, 20
Alpha Canum Venaticorum, 31
Alpha Cygni, 68, 70, 71
Alpha Fornacis, 11
Alpha Piscis Austrinus, 72
Alpha Scuti, 61
Alpha Serpentis, 52
Alpha Ursae Majoris, 20, 22
Alpha Virginis, 28
Anderson, Jay, 21
Andromeda, 5, 9
Andromeda Galaxy, 4, 5, 6, 39, 74
Antares, 53
Aquarius, 72
Aquila, 63
Aratus, 21
Arcturus, 51
Aristotle, 68
Atlas, 12
Auriga, 16

B

Barclay, Jim, 19, 21, 32, 44
Barnard 33, 14
Barnard 72, 55
Barnard 87, 58
Barnard 92, 4, 59
Barnard 93, 59
Barns, C. E., 16
Beehive Cluster, 21
Berlin Observatory, 22
Beta Cassiopeiae, 74, 75
Beta Ceti, 11
Beta Canum Venaticorum, 28
Beta Comae Berenices, 50
Beta Crucis, 30
Beta Cygni, 64, 65
Beta Ophiuchi, 58
Beta Doradus, 14
Beta Persei, 11
Beta Scuti, 62
Beta Tauri, 16
Beta Ursae Majoris, 20, 24
Beta[1] Lyrae, 63
Beta[2] Lyrae, 63
Betelgeuse, 16
Big Dipper, 20, 22, 24, 26, 28, 31, 49, 51
Binary stars, 4, 21
Binoculars, 4, 8, 12, 13, 29, 31, 54, 64, 65, 68
Black holes, 23, 32
Blackeye Galaxy, 30
Blinking Planetary, 64
BM Scorpii, 55
Bode, Johann E., 22, 54
Bootes, 51
Bright nebulae, 4
Brownlee, Alexander K., 25
Bubble Nebula, 74
Butterfly Cluster, 55

C

California Nebula, 12
Camelopardalis, 20
Cancer, 21
Canes Venatici, 26, 28, 31, 49, 50, 52
Canis Major, 19
Canopus, 14
Carina, 21, 23, 24
Cassiopeia, 8, 74, 75
Cataclysmic variables, 70
CBS television logo, 23
Centaurus A, 4, 32, 37, 50
Cepheus, 70
Cerro Tololo Interamerican Observatory, 8
Cetus, 10
Checkmark Nebula, 61
Cheseaux, Phillippe, de, 61
Chi Ursae Majoris, 26
Christmas Tree Cluster, 18
Coal Sack, 29, 30
Coco, Mark J., 38
Cocoon Nebula, 43, 71
Coma Berenices, 25, 26, 27, 30
Coma Star Cluster, 27, 50
Cone Nebula, 18
Coombs, Lee C., 8, 25, 52, 54, 55, 68
Corona Borealis, 52
Corvus, 28
Crab Nebula, 4, 13
Crescent Nebula, 65
Cygnus, 64, 65, 66, 68, 70, 71

D

Dark nebulae, 4, 14, 18, 29, 55, 58, 59
Delta Aquilae, 63
Delta Cancri, 21
Delta Corui, 28
Delta Cygni, 64
Delta Doradus, 14
Delta Geminorum, 19
Delta Sagittae, 64
Delta Tauri, 13
"Demon Star," 11
Deneb, 68, 70, 71
Denebola, 26
Dilsizian, Rick, 12, 53
Dorado, 6, 14
Double Cluster, 10, 35
Double stars, 4, 9, 19, 20, 21, 22, 27, 28, 30, 31, 50, 51, 58, 64, 66, 70
Dragesco, Jean, 9
Dubhe, 20, 22
Dumbbell Nebula, 4, 65
Dwarf novae, 70
Dyer, Alan, 27

E

Eagle Nebula, 38, 61
Eclipsing binary stars, 11
18 Canum Venticorum, 31
84 Ursae Majoris, 51
81 Ursae Majoris, 51
86 Ursae Majoris, 51
83 Ursae Majoris, 51
Electra, 12
Elliptical galaxies, 5
Emission and reflection nebulae, 56
Emission nebulae, 4, 12, 14, 16, 18, 24, 56, 61, 68, 70, 71, 74
Enif, 68
Epsilon Carinae, 21
Epsilon Cygni, 66
Epsilon Herculis, 53
Epsilon Pegasi, 68
Epsilon Piscis Austrinus, 72
Epsilon Tauri, 13
Ernest, Orien, 52
Eskimo Nebula, 4, 19
Espenak, Fred, 68
Eta Cancri, 21
Eta Carinae Nebula, 24, 44
Eta Corvi, 28
Eta Geminorum, 16
Eta Herculis, 53
Eta Pegasi, 74
Eta Tauri, 12
Eta Ursae Majoris, 4-9

F

Fabricius, David, 10
Farrell, J. A., 13
52 Cygni, 66
5 Serpentis, 52
Floyd, Chris, 30
Fomalhaut, 72
Fornax, 11
41 Comae Berenices, 50
47 Aquarii, 72
47 Tucanae, 4, 5, 6, 8
4 Cassiopeiae, 74
14 Comae Berenices, 27
14 Vulpeculae, 65

G

G Doradus, 14
Galactic equator, 55
Galaxies, 4
Gamma Andromedae, 4, 9
Gamma Cancri, 21
Gamma Canis Majoris, 20
Gamma Cygni, 65
Gamma Lyrae, 63
Gamma Sagittarii, 56, 58
Gamma Scuti, 61
Gamma Tauri, 13
Gamma Ursae Majoris, 26
Gamma Virginis, 28
GC 24537, 56
Gemini, 16, 19
Germano, Martin C. 10, 11, 16, 18, 19, 20, 22, 26, 28, 50, 54, 55, 58, 59, 63, 64, 66, 67
Ghost of Jupiter, 23
Globular star clusters, 4, 8, 32, 53, 50, 52, 54, 62, 64, 68
Great Square of Pegasus, 74
Green, Allan, 24

H

h4027, 21
h4031, 21
Hallas, Tony, 5, 15, 17, 33, 34, 39, 41, 71
Halley, Edmond, 24
Harvard College Observatory, 6, 70
Healy, David, 6, 57, 59
Helical Nebula, 72
Helium burning, 10
Hercules Cluster, 4, 50, 53, 62
Hercules, 53, 54
Herschel's Garnet Star, 70
Herschel, John, 30
Herschel, William, 21, 51, 58, 61
Hipparchus, 21
Hooley, Mace, 35
Horsehead Nebula, 14, 41
Horseshoe Nebula, 61
Hubble's Variable Nebula, 18
Hyades, 13
Hydra, 23, 50
Hydrogen burning, 10

I

I1104, 21
Iburg, Bill, 20, 43
IC 434, 14
IC 1396, 70
IC 2602, 23
IC 4665, 58
IC 5067-70, 68
IC 5146, 71
Ihle, Abraham, 62
Irregular galaxies, 23
Irregular variable stars, 52

J

Jewel Box Cluster, 30
Jupiter, 4
Justis, Preston S., 14, 74, 75

K

Kappa Tauri, 13
Karkoschka, Erich, 4
Kirch, Gottfried, 62
Kitt Peak National Observatory, 6

L

Lagoon Nebula, 4, 33, 56, 58
Lambda Leonis, 22
Lambda Sagittarii, 62
Lambda Scorpii, 55
Lambert, Phillip J., 40
Large Magellanic Cloud, 6, 14, 43
Lenticular galaxies, 32
Leo, 22, 25, 26
Lilge, Alfred, 12, 53, 65, 66, 70
Little Dumbbell Nebula, 9
Long period variable star, 10
Lovi, George, 4
Lyra, 63

M

M1, 13
M3, 50, 52
M4, 4, 53
M5, 52
M6, 55
M7, 55, 55
M8, 56, 58
M11, 62
M13, 53, 54, 62
M15, 68
M16, 61
M17, 61
M20, 56, 58
M21, 56
M22, 4, 62
M24, 59
M27, 9, 65
M31, 5
M32, 5
M33, 4, 8, 42
M35, 16
M37, 16
M39, 68
M41, 19
M42, 14, 24
M44, 21
M45, 12
M46, 55
M51, 49
M52, 74
M57, 63
M63, 31
M64, 30
M65, 25
M66, 25
M71, 4, 64
M76, 9
M81, 4, 22
M82, 22, 23
M83, 50
M92, 54
M94, 28
M97, 24
M98, 25
M99, 25, 26
M100, 26
M101, 4, 34, 51
M104, 28
M106, 26
M108, 24
Magellanic Clouds, 5, 6, 14
Maia, 12
Marling, Jack B., 23, 37, 46
Mars, 4
McGlaun, Daniel, J. Jr., 51
Mechain, Pierre, 25
Melotte 111, 27
Merek, 20, 24
Melotte 25, 13
Melotte 22, 12
Merope, 12
Messier, Charles, 13, 16, 22, 55, 59

Milky Way, 4, 5, 6, 8, 21, 27, 29, 38, 53, 56, 59, 68, 70, 71
Miller, Ron, 9
Mimosa, 30
Mira, 4, 10, 11
Mizar, 4, 31, 51
Monoceros, 16, 18
Montanari, Geminiano, 11
Moon, 4, 12, 14, 16, 19, 21, 32, 51, 54, 55, 68, 71
Mount Daphne, 5, 15, 17, 33, 34, 39, 41, 71
Mu Centauri, 32
Mu Cephei, 70
Mu Geminorum, 16
Mu Hydrae, 23
Multiple stars, 4
Musea, 29
Myers, Larry, 42

N

Nebula filters, 12, 19, 56, 65, 68, 72
Newton, Jack, 19, 46, 48, 62, 64, 69
NGC 205, 5
NGC 206, 5
NGC 209, 16
NGC 224, 5
NGC 253, 6, 46
NGC 362, 6, 8
NGC 457, 8
NGC 598, 8
NGC 650-1, 9
NGC 752, 9
NGC 869, 10
NGC 884, 10
NGC 1360, 4, 11
NGC 1435, 12
NGC 1499, 12
NGC 1952, 13
NGC 1976, 14
NGC 2070, 6, 14
NGC 2158, 16
NGC 2168, 16
NGC 2237-9, 16
NGC 2244, 16
NGC 2261, 18
NGC 2264, 18
NGC 2287, 19
NGC 2392, 19
NGC 2403, 20
NGC 2437, 20
NGC 2438, 20
NGC 2516, 21
NGC 2632, 21
NGC 2841, 22
NGC 2903, 22
NGC 3031, 23
NGC 3034, 23
NGC 3242, 23
NGC 3372, 24
NGC 3587, 24
NGC 3623, 25
NGC 3627, 25
NGC 3628, 25
NGC 4254, 25
NGC 4258, 26
NGC 4321, 26

NGC 4565, 27
NGC 4594, 28
NGC 4736, 28
NGC 4755, 30
NGC 4826, 30
NGC 5055, 30
NGC 5128, 32
NGC 5139, 32
NGC 5194, 49
NGC 5195, 49
NGC 5236, 50
NGC 5272, 50
NGC 5457, 51
NGC 5904, 52
NGC 6121, 53
NGC 6205, 53
NGC 6207, 53
NGC 6231, 54
NGC 6341, 54
NGC 6405, 55
NGC 6475, 55
NGC 6514, 56
NGC 6523, 56
NGC 6530, 56
NGC 6603, 59
NGC 6611, 61
NGC 6618, 61
NGC 6656, 62
NGC 6664, 61
NGC 6705, 62
NGC 6720, 63
NGC 6781, 4, 63
NGC 6826, 64
NGC 6853, 65
NGC 6888, 65, 66
NGC 6960, 66
NGC 6979, 66
NGC 6992, 66
NGC 6995, 66
NGC 6997, 68
NGC 7000, 68
NGC 7078, 68
NGC 7092, 68
NGC 7293, 72
NGC 7317, 74
NGC 7318A, 74
NGC 7318B, 74
NGC 7319, 74
NGC 7320, 74
NGC 7331, 74
NGC 7538, 74
NGC 7635, 74
NGC 7789, 75
19 Canum Venaticorum, 31
90 Tauri, 13
North America Nebula, 36, 68
Norton's 2000.0, 4
Nu Aquarii, 72
Nu Centauri, 32

O

Omega Centauri, 4, 5, 32 46
Omega Nebula, 4, 38, 61
Omicron Ceti, 10
Omicron Ursae Majoris, 20
1 Geminorum, 16
Open star clusters, 4, 8, 10, 12, 13, 16, 19, 20, 21, 23, 27, 30, 54, 55, 56, 62, 68, 74, 75

Ophiuchus, 55 58
Optical doubles, 4
Orion Nebula, 4, 14, 24, 34
Orion, 14
Orion's sword, 14
Owen, Todd, 62, 70
Owl Cluster, 8
Owl Nebula, 24

P

Parrot's Head Nebula, 58
Parsons, William, 27
Pegasus, 68, 74
Pelican Nebula, 68
Perseus, 9, 10, 11, 12
Pfeiffer, Randall R., 34
Phecda, 26
Phi Cassiopeiae, 8
Pi Herculis, 53, 54
Pi Hydrae, 50
Pi^2 Cygni, 71
Pickering, Edward C., 14
Pigott, E., 52, 62
Pinwheel Galaxy, 8
Planetary nebulae, 4, 9, 11, 19, 20, 23, 24, 63, 64, 65, 72
Pleiades, 12, 13, 40
Pleione, 12
Pluto, 31
Porrima, 28
Praesepe, 21
Provin, Robert, 10, 11, 21, 31, 58
Psi Centauri, 32
Ptolemy, 55
Pulsation theory, 10
Puppis, 20

Q

Quirk, Steve, 14

R

R Coronae Borealis, 4, 52
R Monocrerotis, 18
R Scuti, 4, 62
RZ Fornacis, 11
Rappaport, Barry, 4
Reflection nebulae, 4, 18
Rho Cassiopeiae, 75
Rho Cygni, 68, 70, 71
Riccioli, Giovanni, 31
Ridpath, Ian, 4
Ring Nebula, 4, 11, 63
Roberts, Issac, 22
Roques, Paul, 18, 22, 31
Rosette Nebula, 4, 18, 44
Rosse, Third Earl of, 27
Royer, Ronald, 7, 29
RV Tauri-type variable stars, 62

S

S Nebula, 4, 47, 55
Sadr, 65
Sagitta, 64
Sagittarius, 53, 55, 56, 58, 59,

61, 62
Sanford, John, 28, 64
Saturn, 4
Scorpius, 53, 54, 55
Sculptor, 6
Scutum, 62
Scutum star cloud, 62
Serpens, 52, 61
Seven Sisters, 12
70 Ophiuchi, 58
75 Cygni, 70
71 Comae Berenices, 27
71 Tauri, 13
Shaula, 55
Sickle, 22
Σ 1386, 22
Σ 1387, 22
Sigma Cassiopeiae, 75
Simmons, Mike, 44
Sirius B, 19
Sirius, 20
6 Comae Berenices, 25, 26
6 Serpentis, 52
16 Comae Berenices, 27
16 Cygni, 64
68 Tauri, 13
63 Ursae Majoris, 20, 22
Sky Atlas 2000.0, 4
Small Magellanic Cloud, 5, 6, 8
Small Sagittarius star cloud, 59
Smyth, William H., 16, 22, 61, 68
Snake Nebula, 55
Sollee, Neyle, 13
Sombrero Galaxy, 4, 28
Southern Cross, 23, 29
Southern Pleiades, 23
Spica, 28
Spiral galaxies, 5, 6, 8, 20, 22, 25, 26, 27, 28, 30, 31, 49, 50, 51, 53, 74
"Spiral nebulae," 27
SS Cygni, 4, 70
Star cloud, 59
Stecker, Michael, 47
Stephen's Quintet, 74
Sun, 10, 19, 21, 28, 31, 62
Supergiants, 30
Supernovae, 26
Supernovae remnants, 13, 65, 66
Surface brightness, 12, 26, 28, 49, 50, 51, 61
Swan Nebula, 61

T

Table of Scorpius, 54
Tarantula Nebula, 6, 14
Taurus, 12, 13
Taygeta, 12
Teapot, 56, 59, 62
10 Serpentis, 52
The Observer's Sky Atlas, 4
Theta Cancri, 21
Theta Carinae, 23
Theta Leonis, 25
Theta Ophiuchi, 55
$Theta^1$ Orionis, 14

$Theta^1$ Tauri, 13
$Theta^2$ Orionis, 14
$Theta^2$ Tauri, 13
13 Comae Berenices, 27
13 Monocerotis, 16, 18
13 Vulpeculae, 65
30 Comae Berenices, 27, 50
35 Comae Berenices, 30
31 Comae Berenices, 50
37 Lyncis, 22
Thompson, Gregg D., 23
Tirion, Wil, 4
Trapezium, 14
Triangulum, 8
Trifid Nebula, 33, 56, 58
Triple stars, 21, 56
Tsang, Simon, 43
Tucana, 6, 8
12 Comae Berenices, 27
20 Canum Venaticorum, 31
24 Ursae Majoris, 22, 49
21 Comae Berenices, 27
26 Aurigae, 16
23 Canum Venaticorum, 31
22 Aquilae, 63
22 Comae Berenices, 27
Tyree, Vance C., 36

U

U Geminorum-type variable stars, 70
Ursa Major, 51

V

Variable stars, 4
Veil Nebula, 48, 66
Virgo, 28
Virgo Cluster, 26, 28
Vogel, H. C., 11
Vulpecula, 9, 65

W

Wallis, Brad, 11, 21, 31, 58
Whirlpool Galaxy, 4, 34, 49
White dwarfs, 19
Wild Duck Cluster, 62
Wilson, Matthew, 38
"Wonderful, The," 10

X

Xi Bootis, 4, 51
Xi Persei, 12

Z

Zeta Cancri, 4
Zeta Centauri, 32
Zeta Herculis, 53
Zeta Orionis, 14
Zeta Tauri, 13
Zeta Ursae Majoris, 31
$Zeta^1$ Scorpii, 54
$Zeta^2$ Scorpii, 54
Zussman, Kim, 8, 9, 22, 23, 24, 26, 27, 28, 30, 49, 51, 56, 60, 61, 62, 63, 65, 73, 74

Bibliography

ASTRONOMY. Kalmbach Publishing Co., Waukesha, Wisconsin. Founded in 1973, this monthly is the largest English-language astronomy periodical. It regularly contains plentiful information about deep-sky objects.

Berry, Richard. *Discover the Stars.* 119 pp., paper. Harmony Books, New York, 1987. The former Editor-in-Chief of ASTRONOMY introduces naked-eye, binocular, and small telescope observing using twelve all-sky maps and twenty-three close-up maps.

Burnham, Robert. *The Star Book.* 17 pp., spiral-bound. AstroMedia and Cambridge University Press, Milwaukee, 1983. Group of seasonal star maps from ASTRONOMY magazine with a brief discussion of what to see.

Burnham, Robert Jr. *Burnham's Celestial Handbook.* Three vols., 2,138 pp., paper. Dover Publications, New York, 1978. This voluminous compilation of deep-sky objects contains many photographs, charts, and tables.

Consolmagno, Guy, and Dan M. Davis. *Turn Left at Orion: a Hundred Night Sky Objects to See in a Small Telescope — and How to Find Them.* 205 pp., hardcover. Cambridge University Press, New York, 1990. A brief description, sketch, and finder chart for one hundred objects of unusual interest to amateur observers.

Dixon, Robert S., and George Sonneborn, compilers. *A Master List of Nonstellar Optical Astronomical Objects.* 835 pp., hardcover. Ohio State University Press, Columbus, 1980. An all-in-one compilation of more than 185,000 deep-sky objects drawn from 270 catalogues. This list is invaluable for identifying objects.

Dreyer, John Louis Emil. *New General Catalogue of Nebulae and Clusters of Stars* (1888). *Index Catalogue* (1895). *Second Index Catalogue* (1908). 378 pp., paper. Royal Astronomical Society, London, 1962. This single volume presents a facsimile reprint of the original basic listings of deep-sky objects compiled by Dreyer.

Eicher, David J., and the editors of Deep Sky magazine. *Deep Sky Observing with Small Telescopes.* 331 pp., paper. Enslow Publishers, Hillside, New Jersey, 1989. A beginner's manual for observing deep-sky objects with 2-inch to 6-inch telescopes. Contains extensive listings of objects and many photographs and eyepiece sketches made by backyard observers.

Eicher, David J. *The Universe from Your Backyard.* 188 pp., hardcover. Cambridge University Press and AstroMedia, a division of Kalmbach Publishing Co., New York, 1988. This book is a series of republished "Backyard Astronomer" articles from ASTRONOMY magazine. Included in its coverage are forty-six constellations or groups of constellations and 690 deep-sky objects. A three-color map, eyepiece sketches, and color photographs appear for each constellation.

Hartung, E.J. *Astronomical Objects for Southern Telescopes.* 238 pp., hardcover. Cambridge University Press, New York, 1968. Meticulous observing notes by an Australian observer for many deep-sky objects in the Southern Hemisphere.

Hirshfeld, Alan, and Roger W. Sinnott, eds. *Sky Catalogue 2000.0.* Cambridge University Press and Sky Publishing Corp., New York, 1982-1985. Two vols. Volume two (356 pp., hardcover) lists fundamental data for thousands of double and variable stars, 750 open clusters, 150 globular clusters, 238 bright nebulae, 150 dark nebulae, 564 planetary nebulae, 3,116 galaxies, and 297 quasars.

Jones, Kenneth Glyn. *Messier's Nebulae and Star Clusters.* 480 pp., hardcover. Faber and Faber Ltd., London, 1968. One of England's foremost amateur astronomers presents descriptions and eyepiece drawings for each of the Messier objects.

Karkoschka, Erich. *The Observer's Sky Atlas.* 130 pp., paperback. Springer-Verlag, New York, 1990. A wonderful pocket-sized star atlas showing enough detail to find bright deep-sky objects.

Luginbuhl, Christian B., and Brian A. Skiff. *Observing Handbook and Catalogue of Deep-Sky Objects.* 352 pp., hardcover. Cambridge University Press, New York, 1990. The authoritative single volume for data on deep-sky objects, this book contains information on nearly 2,050 galaxies, nebulae, and clusters.

Newton, Jack, and Philip Teece. *The Guide to Amateur Astronomy.* 327 pp., hardcover. Cambridge University Press, New York, 1988. The best all-around introduction to what amateur astronomy is all about.

Peltier, Leslie C. *Leslie Peltier's Guide to the Stars.* 185 pp., paper. AstroMedia Corp. and Cambridge University Press, Milwaukee, 1986. A basic introduction to observing stars, planets, and deep-sky objects with binoculars.

Sky & Telescope. Sky Publishing Corp., Cambridge, Massachusetts. The oldest astronomy magazine in America, *Sky & Telescope* contains a monthly "Deep Sky Wonders" column written by the experienced observer Walter Scott Houston.

Tirion, Wil, Barry Rappaport, and George Lovi. *Uranometria 2000.0.* Two vols., 473 quarto-sized charts, hardcover. Wilmann-Bell, Inc., Richmond, Virginia, 1987-1988. A minutely detailed, large-scale atlas, *Uranometria 2000.0* shows 332,556 stars down to magnitude 9.5 and many thousands of deep-sky objects.

Tirion, Wil. *Sky Atlas 2000.0.* Twenty-six fold-out folio charts, spiral-bound. Cambridge University Press and Sky Publishing Corp., New York, 1981. A large-scale atlas showing 43,000 stars down to magnitude 8 and 2,500 deep-sky objects in color.

Vehrenberg, Hans. *Atlas of Deep-Sky Splendors.* Third ed. 246 pp., hardcover. Treugesell-Verlag and Sky Publishing Corp., Dusseldorf, 1978. A splendid photographic album that contains images of hundreds of deep-sky objects all reproduced at the same scale for easy comparison.